May the knowledge of your ancestor struggles give you gratitude. May the understanding of their dreams lift your heart.   KJH

Martyr of Loray Mill

Ella May and the 1929 Textile Workers' Strike in Gastonia, North Carolina

KRISTINA HORTON

Tammy,
You have been a partner with me on this road to discovery. I will be forever grateful for you capturing much of it marvelously on film.

With great love and appreciation.

Kristina

McFarland & Company, Inc., Publishers
Jefferson, North Carolina

ISBN 978-0-7864-9964-9 (softcover : acid free paper) ∞
ISBN 978-1-4766-2243-9 (ebook)

LIBRARY OF CONGRESS CATALOGUING DATA ARE AVAILABLE

British Library cataloguing data are available

Front cover: *inset* Ella May, September 1929 (courtesy of
Millican Pictorial History Museum); *background* cover photograph
of the Old Loray/Firestone Mill by Tammy Cantrell

Printed in the United States of America

McFarland & Company, Inc., Publishers
Box 611, Jefferson, North Carolina 28640
www.mcfarlandpub.com

Table of Contents

Introduction

The Loray Mill Strike was one of the major news events of its day. Men and women across the country opened their morning papers to see what had transpired between hundreds of suffering Southern textile mill workers, their revolutionary communist leaders, and the God-fearing community in which a battle of values was being fought. Journalists sprang to Gaston County, North Carolina. One of them witnessed a scrappy young female textile worker with guitar in hand ascend a speaking platform. At once the reporter was "transfixed" by her presence and the music that "bubbled" from her. Other textile workers were drawn in, attentively listening to her "full, throaty voice," chuckling at her attacks on mill bosses, nodding at testimony to hardship. The crowd delighted in how she "gave the songs in mountain style, with an odd sort of yip at the end." Another female striker commented, "Purtiest singing I've ever heard." Others agreed. The balladeer with the magnetic aura in the center of a national controversy was Ella May.[1]

Today I have a pretty clear image of Ella and her story. As her great-granddaughter, I might be expected to have some insight but this was not the case as I began my research. I was first exposed to the existence of Ella May at the age of nine when I stumbled upon a newspaper clipping about her in a family photo album. The article was about how my grandmother Millie received a personal invitation from Pete Seeger to attend a concert. The article revealed that Millie's mother Ella May was also a singer-songwriter and had inspired Pete Seeger.[2]

Years later as a college student I remembered that article when I was given an assignment to write about an ancestor. This is when I began asking questions about my great-grandmother. I wrote a paper at the age of eighteen, followed by an independent study project at the age of twenty-two, and then this manuscript was completed as I reached my late thirties. So this book is not the product of a scholarly endeavor, researched over a matter of months or a few years. My findings developed slowly over time, as did my interpretations of those findings. It took decades of independent research, reflection, and revision to mature this story to what it is today.

Most of my family had limited knowledge of Ella May's involvement in the Loray Mill Strike until the late 1990s, when information starting appearing on the Internet. Prior to this time, family members without personal memories of her sought answers from their elders with poor results. It had been too painful for older generations to discuss and share. Decades after Ella's death, her adult son Albert Wiggins, three at the time of her death, with his own son Victor Wiggins in tow, tried to obtain information from Ella May's aging brother Wesley. Wesley was an eyewitness to much of what happened. Victor recalled, "My dad and I would sit down and try to get Wes to talk about [Ella May] a lot of evenings, and Dad would ask him questions about his mother, but it was like pulling teeth to get Wesley to talk." Rita Jackson, Ella's granddaughter and my great-uncle Clyde's daughter, who grew up in the area and resides now in Bessemer City, recalled being hushed as a child and told "don't ask any questions" when she inquired about her grandmother.[3]

Family was even more tight-lipped when it came to acknowledging Ella May in public. Ella's children saw firsthand how hated and despised their mother was by members of the community. They knew what awful things their children could hear about their grandmother and there was always the chance that they too could be despised by association. Ella May's grandchildren who grew up in the Piedmont were directed not to tell outsiders who their grandmother was. These grandchildren must have felt confusion and uncertainty about their ancestry.

None of Ella's children ever spoke an ill word about her. They felt no shame about their mother's behavior. It wasn't words spoken by the family that cast doubt upon Ella May's character, it was the words that weren't spoken. For the most part, her grandchildren didn't talk about her in their childhood and many of them continued their silence into adulthood.

However, the silence did not represent apathy. Family members felt and feel pain from losing a family member. My own sixty-two-year-old mother Darlene, Ella's granddaughter, can even today well up with emotion when she thinks of the loss. Ella's kin from every generation feel that an indefensible injustice has been done to the family.

The family legacy passed down from Ella May is a complicated one. Fighting for one's rights is valued, for example, but not in an organized way. Ella's daughter Millie forbade anyone in the family from associating with unions. She spoke up against my grandfather Merritt's joining the union that formed at the Westinghouse plant in Horseheads, New York, where he worked. Millie said in an interview, "I didn't want to lose him like I lost my momma." She told my mother Darlene that it would have broken her heart to see any of her own babies walking the picket line.[4]

My grandmother's dislike of unions and painful feelings about the past make me wonder what she would have thought of this book, the discoveries and conclusions I have made. I would hope that her pride would outweigh the pain. My grandmother passed away in December 1986, when I was ten years old. I do not recall any of the stories she told our family of her childhood.

I remember Millie as a frail old woman in poor health. Yet she was a pillar of strength for my mother's large family. Millie had fourteen children and dozens of grandchildren. She was a large gray-haired woman with pinned-back hair who cooked wonderful homemade Southern food and spoke frankly and brashly. As a young child I was intimidated by her, although she never spoke a harsh word to me. I felt like she had the most powerful aura of any person I have ever encountered.

An indescribable inner strength emanates from the women in my family. My mother, sister, aunts, and cousins are all strong women. We were brought up to be tough. I was never coddled as a child. Like my mother, and I imagine my grandmother and great-grandmother, I was a sensitive child. I felt things easily and deeply, but I was taught to react to emotional and physical discomfort, not wallow in it. When I scraped an arm or bruised a knee I was not given hugs and kisses. I was given a bandage and direction on how not to injure myself in the future. I am not a mother, but I am an elementary school teacher, and I find that I treat my students in a similar manner. My mother was more affectionate than her mother, giving us hugs and kisses and saying "I love you" at bedtime. Millie never gave her children those emotional luxuries, although all of her children knew they were loved. Millie would sometimes steal some moments away from domestic work to go into the backyard and relax with her girls. She would spread a blanket under a handful of mature maple trees, lie down, and observe the leaves and drifting clouds. In this relaxed state she shared with her young girls the memories of her own childhood.[5]

The most prominent, enduring family legacy is strength. In an interview, Millie said the following about her mother and herself: "She always told us to stick up for ourselves.... She always fought for what she believed in. I'll stand up, I don't care who it is." In another interview she further explained, "She told me, 'Always stick up for your rights because if you don't no one else will.'" After reading these statements from my grandmother and researching my great-grandmother I now have a better understanding of the source of my family's female grit.[6]

Initially my reaction to the discoveries I made was anger. My great-grandmother was murdered, killed while trying to improve the lives of her children. No one was convicted for her murder, nor was anyone convicted for the numerous other crimes committed against textile strikers that year in the South. Few locals sided with the textile workers, striking to improve their conditions. Well-respected members of the community, businessmen, government officials, and American Legion men, organized against the strikers and their leaders, creating the vigilante group the Committee of 100. The Loray Mill furnished bail to those accused of crimes against union members in the area. Months, even years after the strike, esteemed members of the community spoke ill of union leaders, including Ella May.

As my understanding of the situation in Gaston County grew, my anger lessened. There were reasons why members in the community reacted violently against union members. Survival was a precarious thing in 1929. A community that was brought back to life by the textile industry after the devastation of the Civil War again felt threatened. If the textile industry were to fall due to a drawn-out strike, then so would the communities it supported. Fear, especially fear of destruction, can drive individuals to do radical things. Groups of individuals, a mob such as the Committee of 100, cloaked in the guise of justice can even more easily allow fear to drive the group to extreme behavior. This fear was justified. Northerners had destroyed their way of life before. While justified fear hardly justifies acts of violence, it is something that can be understood.

4

So I now see Ella May's murder as the result of a series of events that avalanched to a horrific outcome.

Decades later, key players in the strike themselves gained insight into the strike that they led. The primary strike leader, Fred Beal, after viewing conditions in Russia, no longer believed the words that he himself had given workers on behalf of the Communist Party. He no longer trusted that an equal and fair workforce could have been developed under the guide of the party's National Textile Workers Union. Topmost female organizer and friend to Ella, Vera Buch, also became disillusioned by the party, but was optimistic that the organizers' intent made a positive impact in the future. Vera reflected on the cause in a 1970s magazine article:

> We felt that had we only more funds and organizers to extend the strike, including other mills and other companies, we might possibly win a victory. This today I must question. Our limitations were part of the historical picture, which included the opposition of the AFL and the incompetence of the Communist Party and the treachery of its leaders as well as the opposition of the employers and of the government. In any case, a lost strike is not really lost; if conducted militantly, honestly and effectively, it lays the basis for future battles. If no resistance is attempted against such ferocious exploitation as these textile masters enforced and still do, only hopeless slavery can result. I feel we contributed our little bit to the long struggle for freedom.[7]

For many years I was unable to see this wedge of progress that Ella May, the other strikers, and their leaders made. Vera believed at the end of her life that the Loray Mill Strike inspired others to make headway. I did not initially see things that way. The plight of Southern textile workers got worse, and stayed that way for years before it slowly started to improve. It appeared to me that Ella and her children suffered hatred and violence with nothing to show for it. However, over the years I've gained an understanding that it is not just those who actually see change from their immediate actions that are solely responsible for the social progress that later takes place. Hundreds, perhaps thousands, can stand up against the status quo for years before the status quo is altered. These countless individuals who brave to pioneer change are integral to the change that develops years later. Groundbreaking activists open the minds and rouse courage in those who follow them.

Ella May inspired those who sought change immediately following her death and she motivates others to this day. I've read a great number of articles, books, songs, and plays that talk about Ella May's story, that share her hope. Some of these historical accounts were constructed in the 1930s, and others are more recent creations. Through my journey of discovery I have met and heard from countless individuals who are deeply touched by her struggle. Reading about and seeing firsthand how moved others are by Ella May's story reinforced that her work and sacrifice weren't in vain. None of the strikers' efforts were wasted, not as long as they are remembered and behavior has been changed in some way as a result.

What does that mean to me, the researcher, a great-granddaughter? Ella's ability to speak the truth and instigate change in the face of the worst kind of adversity motivates me to try to do the same. I am known for speaking up when I feel an injustice has been done. I've attended protest rallies organized against corporate greed and that promote social justice. I stand up for what I believe is right for myself, my peers, and most importantly, my students. I feel a great sense of duty to children, and to Ella May. Being the best educator I can be and sharing Ella May's story are the two biggest priorities in my life. I am sometimes accused of spending too much time on what I do. However, I make no apologies for trying my best at teaching or for spending thousands of hours on research and writing. What has been given to me, my life, my opportunities, have not been given so easily. Pain has been suffered and blood has been shed for what I have. Giving to the next generation through my teaching and my research is an attempt to pay back a bit of the ancestral gifts that have been given to me.

Ella May was my mother's mother's mother. I know little about my maternal family's history firsthand from my grandmother. One story I do remember from childhood involves my grandmother's brother, Clyde, and me. The version I share is not the most factual account (family stories change over time), but my personal recollection of the story.

I was born on March 25, 1976, which happened to be the same day as my great-uncle Clyde's funeral. My grandmother Millie, as well as other family members, were actually at the services being held in Gastonia, North Carolina, as I was being born in Pennsylvania. The family was somber, as would be expected. My grandmother was especially heartbroken, for the memories of her childhood and the harsh life she and her brother had shared were reawakened. Clyde was closest in age to Millie, being only two years older. Their experiences and views of what had taken place in their early childhood in Gaston County must have been quite similar.

My grandmother was a tough woman like her mother, but the grief and pain were visible nonetheless on her strong face. She, like the others, took her turn to view her brother for one last time. As Millie looked down to see Clyde she saw something that the others had not. Her expression suddenly changed from sadness to anger. She demanded to know in a harsh fiery voice: "Who put that thing on his body? Who did that? Why is that there?" The quiet room suddenly became even quieter. Bewildered family members didn't know what to say or do. Millie continued yelling and demanded answers.

Finally, a relative stepped up to Millie and asked her what she was talking about. Without hesitation she pointed to her brother's forehead and identified "that." Still baffled, the man asked "What? I don't see anything." Millie questioned, "You don't see that star on his head? You mean you don't see that white star?" Eventually my grandmother was calmed and reassured that no white star was painted on her brother's forehead. A few moments after the ordeal, while everyone was still recovering from the outburst, my Aunt Candy entered the room. She announced to everyone that my mother Darlene had just given birth to a healthy baby girl. The only thing distinctive about the child was that she had a white spot on her forehead.

Well, the spot went away, but the story stayed with me. Other family members remember various versions of this same tale, with the main points of the white star on Clyde's forehead and the white spot on my forehead remaining the same. What is the significance and relevance? To Millie, the star was a sign of hope, that for everything taken away, something is given in return. In this case her brother was taken and a grandchild was given. My uncles teased me as a child saying I was their uncle reincarnated. I did some of my own research and found that in Christianity a white star on a body symbolizes enlightenment, one who gives knowledge to others. It very well may all be a coincidence, but I like to think that I am doing that very thing. I'm trying to enlighten the world about what happened to Clyde and the other children, about the story of Ella May, about the Loray Mill Strike, and about our textile history.

Introduction

As I mentioned earlier, I am no historian, sociologist, or accomplished writer. I began the research involved in the creation of this book in order to find answers for myself, to understand my own roots, my own history, in order to attempt to comprehend the how and why of my great-grandmother's tragedy. I discovered more questions, finding a wider scope of issues involved in the story than I originally anticipated. As I progressed throughout the research I began to piece together and create a holistic, clearer picture for myself, incorporating a variety of perspectives. I looked at the direct events that occurred in 1929, as well as those precursing Ella May's life, and those following her death. Early on I realized how valuable this story I reconstructed is, not only to me, but to everyone. This is not only part of my history, but it is part of our communal history.

In 1998 I began searching for the whereabouts of the mill where Ella May worked, American Mill No. 2. Unlike the majority of seniors in college on spring break, my best friend Christine Bowman and I did not head to the beaches. We headed to the heart of the Piedmont, Gaston County, to find answers. While researching in the Bessemer City Library I was told that the American Mill No. 2 had been torn down. For years I took this piece of information as truth, having never run across any other bits of information about it. It seemed as if the mill indeed had disappeared. However, in 2013 I was surprised to discover the American Mill No. 2 still stood.

Former Bessemer City resident, Tammy Cantrell, volunteered to help me find an image of the American Mill No. 2. She discovered some books documenting the history of the city. No images of the American Mill No. 2 were immediately apparent. However, with some digging together we made connections between the former mill names and management to discover that pictures of the mill did indeed exist. Even more importantly and shockingly, Tammy recognized the last of the long list of names the mill has donned. Dawn Processing long ago was the American Mill No. 2, and Tammy knew where Dawn Processing still stood. Before long I had the privilege of standing in the last textile mill Ella May had stood in.

Joe Depriest, staff writer of the *Charlotte Observer*, interviewed me in February 2004. He discovered my search for Ella May's story while I and others were preparing for the seventy-fifth anniversary symposium.

The resulting article went to the Associated Press. Soon a variety of supporters contacted me. Local Piedmont residents called and shared their own small bits of personal or family recollections of the Loray Mill Strike. North Carolina natives with experience researching local history, such as George Loveland and Roxanne Newton, offered generous research and moral support. Crystal Lee Jordan Sutton, the woman the movie *Norma Rae* was based upon, called and encouraged me. Loyal and Garnett Jones, brothers of Troy Jones, one of the men involved in the murder of Ella May, shared their own personal information about the events leading up to and following her murder. The widespread outside interest opened my eyes to the importance others already felt this piece of history holds. It deepened my resolve and sense of duty towards this project.

A key piece of information that I wanted to discover early on was who actually shot Ella May. This was difficult because most sources were uncertain. After digging into 1929 and 1930 editions of the *Gastonia Gazette*, I discovered the shooter was most likely an employee of the Loray Mill named Horace Wheelus. It is also highly likely that he was given the day off and paid to hamper the strikers' efforts, along with a number of Loray Mill employees.[8]

Not Wheelus, but another man, reportedly confessed to the murder of Ella May. The man known as Old Man McEntire confided in some friends that Ella confronted him and pulled out his hair in his dreams. This began shortly after Ella's death and continued up until his own demise. McEntire suffered from poor health, including hair loss. He died from his ailments a few years after Ella May's murder. McEntire reportedly confessed to the murder on his deathbed in a Gaston County hospital. He called a judge, an undertaker, and a minister to his bedside. He asked, "If a man shoots and kills a pregnant woman, is he guilty of one murder or two?" He asked additional questions similar in nature until he was confronted about why he was asking them. His reply was, "Because I killed Ella May Wiggins." He died soon after. He was buried in the same Bessemer City cemetery as Ella May, near her grave. Decades later on a fall afternoon, I spotted Thomas McEntire's headstone near Ella May's. He died in 1932, and inscribed below his name was "President Southern Textile Association." I later discovered that Ella May's daughters were told the same story by locals and that Thomas McEntire was identified as the confessor. So indeed the grave that I visited that clear chilly, day belonged to a man who believed he murdered my great-grandmother.[9]

There were a number of false confessions. Frank Sisk, the undertaker who guarded Ella's body and her children, also was confessed to by a man on his deathbed. John R. Mason, a millworker, revealed to Sisk that he was responsible for Ella May's untimely demise. A number of men were shooting that day. The chaos may have led to a number of guilty consciences.

Author Robert Williams chose not to reveal the name of the confessor he heard about in his book, *The Thirteenth Juror*. He explained his decision:

> The guilty man was not, as many people suggested, a high-ranking official of the Loray mill. He was not connected officially with the Committee of One Hundred, and he was never linked in life in any way to the murder. In short, nothing in the man's public life would have suggested that he would have been the perpetrator of one of the most heinous crimes in area history. Those who knew him found him to be a quiet man, if opinionated, one who was devoted to his family and friends. Rumors that the man was hired to do the shooting have never been substantiated.
>
> The identity of the man is withheld here for two reasons: first, he is deceased and is unable to defend himself against charges and at the same time he is beyond the reach of the law; second, there seems to be little reason to bring public embarrassment, humiliation, and personal emotional damage to friends and family members who survive. The story provided by the witness has been related to the Gastonia chief of police, who agrees that nothing of a positive nature could be derived from a public revelation of the identity of the man.
>
> There is a third reason: the man's grave is, even so long after his death, marked by fresh flowers regularly, and these flowers, symbolizing the feelings of persons still living in the area, argue far more eloquently for privacy than an ill-directed desire for sensationalism could argue for public announcements.[10]

I disagree with Williams' sentiment that disclosing the painful truth is the wrong thing to do. This way of thinking has kept the Loray Mill Strike story from being shared with family members. It is why descendants of those involved know so little about their families' history. Williams clearly is not alone in a wish to keep some painful facts a secret. However, he published his reasons for not revealing the murderer of my great-grandmother.

His words angered me when I read them, and I was still disturbed by his response when I met the author when I was in my twenties. In 2004, I confronted an off-guard Williams during one of his many presentations about Gaston County history at the Gaston County Library. After he discussed the Loray Mill Strike, he took questions from the audience. I raised my hand, explained that I was a great-granddaughter of Ella May, and asked him why he chose not to reveal who he believed murdered her, and about Ella May's family's right to know the truth. No doubt these were not the types of comments or questions he was expecting. He reiterated the reasoning he had previously published; his words were gentle, almost apologetic. He promised to share with me in private who he believed the murderer was at the end of his presentation, and he did: Old Man McEntire. I appreciated his revealing the name to me, although based on my research I do not believe he is the one that fired the fatal shot. I now feel no bitterness towards his decision not to publicly share, and somewhat bad for putting the man on the spot, but still believe the truth, the whole truth, is important to have.

Many who lived through the painful events of 1929 in the Piedmont, and especially in Gastonia, found the events too bitter and raw to face altogether. This was felt by both mill supporters and strike supporters alike. The local textile documentary *Spinning Through Time* explained that the events of that year created a "black eye" for the area's image and a "bitter pill" for the community to swallow. Friendly, warm, polite Southern culture further promoted the hushed state of the affair long after personal individual families' wounds had healed. Descendants of key players in the 1929 strike carried prominent surnames that have contributed much to the local community over the years. Their names to this day are still very much visible and viable in the community. No one wished to tarnish those respected individuals by associating them with unsavory happenings of the past.[11]

In the last decade or so, the tide turned. Curiosity now overrides a fear of shame. Members of the community aware of the Loray Mill Strike are now hungry to know more details. In April of 2004, I attended an informational session about the strike at the Gaston County Public Library with scores of others. Dennis Aderholt, great-nephew of fallen Chief Aderholt, was also at the event. We were treated like celebrities in the packed meeting room. I felt somewhat awkward, undeserving of the attention, yet proud at the same time. It was a bit of a surreal moment when we met and others surrounded us, taking pictures of two descendants of those slain on opposing sides of the Loray Mill Strike. For those community members who witnessed the event, surely it struck them as if history had come full circle. I feel after completing this project that my own journey of discovery has also come full circle. Through many years of research a naïve curious child grew into a reflective adult. Over time revelations were made, found in historical books, in newspaper articles, in the memories of those who witnessed, and in accounts passed down to future generations. That knowledge is now preserved in this book.

While the idea of sharing Ella May's story is something that I've been focused on for many years, the idea of receiving proceeds from retelling Ella May's tragedy does not sit well with me. I also feel somewhat uncomfortable with the level of notoriety that comes with me doing this project. On one hand, publicity is good; it gets the story out. On the other hand, I am given credit and attention for retelling the harsh life of an ancestor. I have done nothing to become a descendent of Ella May. I'm not unique; there are about a hundred of us. In fact, not until the end of this project did I include myself and my journey as part of the story. I remember a few years ago reading my own name in a scholarly journal article about Ella May and feeling rather awkward about it. However, it was then that I realized that I am part of the aftermath. Whether I am comfortable with it or not, I am part of my own tale of discovery.

One thing I was hoping to gain from this project was a sense of closure, a sense of accomplishment in knowing the ordeal Ella and her children suffered through resulted in some improvement. I wanted an outcome more tangible and direct than inspiration. Ella May sacrificed her life in hope of a better life for her children, but with very limited results. Traces of harsh life are still apparent in the lives of much of my family. I hope that the story of Ella's life will aid in some small way some of her descendants. All profits from this book will go towards an educational fund for descendants of Ella May. Ella May enthusiast George Loveland has generously and graciously agreed to give ⅓ of his proceeds from his book *For Our Little Children: Growing Up in the Shadow of the Loray Mill Strike* towards an Ella May educational fund. *For Our Little Children* details the quest of Gastonia native George Loveland to discover the causes and impacts of the local concealment of the Loray Mill Strike. George interviewed me as well as Dennis Aderholt for his book. He interwove experiences from descendants with events during the 1929 Loray Mill Strike. George mixed fact with realistic fiction in his detailed imagined conversations and vivid description of the strike.

A series of exciting events developed in the late summer and early fall of 2011. I read a *Gaston Gazette* article about an Ella May Wiggins Memorial Committee that had organized with the purpose of erecting a monument in Ella May's honor. I read the list of over half a dozen members on the committee and recognized Lucy Penegar's and Roxanne Newton's names. Both were individuals I had been in contact with. I was somewhat disappointed about not being informed directly about the project, but was more stirred by the group's existence. I called both women and discovered that the next day a gathering of family members at Ella's grave was scheduled. I met Kimberley Hallas, the then chair of the committee. We connected and my journey with the Ella May Wiggins Memorial Committee began.

The primary function of the committee, erecting a monument, I remember contemplating years ago as an eighteen-year-old college student. I had found an article about a Gastonia mayor denying a marker to be placed that recognized her murder. I still have the letter I wrote to persuade the current mayor of the time, those many years ago, to allow a marker to be placed. I never mailed that letter. Why? Why have I been so hesitant about putting my own ideas about a monument and educating the public into action? Why has it taken me nearly twenty years to write this book? I believe in at least part it is because I am very reflective. I study things, look at them from different perspectives. While there is value in this, often there is even more value in action.

I am very thankful for the opportunity to join forces with and become a member of the Ella May Wiggins Memorial Committee. Since its inception there have been numerous articles printed in the *Gaston Gazette*, which over these many years has shifted from a catalyst of harm to textile strikers to a catalyst of help to textile descendants. Once it demonized union organizers and rallied fears in the community against textile strikers. In recent years, the paper has been instrumental in sharing textile history, advertising events, and getting Ella May's story out. Gastonia businesses, too, have gotten into the spirit. The Bottle Shoppe

Deteriorating Loray Mill in 2011 (photograph by Leo Hohmann).

held a month-long art exhibit showcasing local artists' work tied to the textile industry. Freeman's Pub created an Ella May Wiggins drink and Spindle City Café created an Ella May meatloaf sandwich. A storyteller told Ella May's story and a singer performed "Mill Mother's Lament" at Poor Richard's Book Shoppe in Gastonia. George Loveland also had his book signing at Poor Richard's Book Shoppe, and I sang "Mill Mother's Lament" at that time. Community textile poetry was shared at Poor Richard's too. This event was especially close to my heart, since I have written poetry since I was a young girl. I like to think I get this inclination to express myself this way from Ella May.

Outings to historic spots in the community have been made by Ella May Wiggins Memorial members. Local historical preservationist and Ella May Wiggins Memorial member Lucy Penegar took a number of us through a tour of the Loray Mill. The heaviness of the past could be felt in the peeling paint, rusting metal, and sagging wooden beams, while the majesty of the place itself was revealed in the sheer expanse of space, sweeping rows of massive windows, and small rare manufacturing remnants of a bygone era.

Tammy Cantrell and I toured the old American Mill No. 2, now

Dawn Processing. After all these years the mill was still producing textile goods. The mill operated at limited hours, with a skeleton crew, but it was still running. The old No. 2, like the Loray Mill, was deteriorating; only a small section of the mill was still in use. The eighty-two-year-old owner of the mill, Henry Moore, had owned it since the sixties. Like many of his generation he had no wish for special attention, although he did enjoy hearing the history of his mill told. Mr. Moore was struggling to keep the mill in operation, still worked long hours himself, and could only pay his workers a little over minimum wage. It appeared that he didn't want a lot of notoriety paid to his mill due to its deteriorated state. However, I was impressed and touched by the fact that the mill was still standing and still in operation, and that there were workers in Ella May's mill.

Whereas the structure of the Loray Mill impressed me most, the hard-working employees of the old No. 2 mill struck me. They toiled daily for very little reward, yet did it diligently. The aging crew was led by a hands-on mill owner. Mr. Moore still rose early every morning and worked every day, because that is what he knew and that is what gave him pride. He contrasted sharply from descriptions of his predecessors, the Goldbergs, who saw mill ownership as a short-term money-making investment. Ownership for Mr. Moore was a way of life, a way to be independent and help others.

Seemingly appropriate, a symbol of fading pride in American craftsmanship lay just outside the mill. The neighboring water tower displayed a made-in-America symbol. Once a proud rallying call, now a deteriorating reminder of days gone by, much like the deteriorating American Mill No. 2.

Guided by the information given to me by Bobbly Bolch, Ella's grandson, Myrtle's son, I led a group of Ella May Wiggins Memorial members to the porch where Ella May's body was laid immediately after her murder. It felt like a sacred place to me. However, the response we received from the owner was shockingly aggressive. There were four of us in the car. Kimberly Hallas, with her recently broken leg, hobbled to the medium-sized white house with its columned porch while the rest of us remained in the car. She was courteous and professional, explaining who we were and our intentions, which was to photograph the porch. Immediately she was countered with aggression and threats. The home-

The old American Mill No. 2 is still in production as of 2013 (photograph by Tammy Cantrell).

owner declared that he had a gun and a dog, and that we better not have taken any pictures. He ranted about how we were all under surveillance. Kimberly reassured him that we had no intentions of taking pictures without his permission.

He followed her shirtless with gun on hip as she retreated to the car. He continued to barrage her with threats and questions. He wanted to know exactly why we were there. Thinking that he may be more receptive to a relative than a committee member, I stepped out of the car to explain. As calmly and sympathetically as possible, I tried to explain that this was the spot where my great-grandmother was taken after she was killed and that it is important to me and those who are trying to preserve history. He said it was nonsense and that he better not hear or see anything about his home in print or on the Internet. It was clear that he feared attention, outsiders invading his home. Encountering this

The "Crafted with Pride in U.S." sign is still faintly visible in 2013 on the water tower adjacent to the old American Mill No. 2 (photograph by Tammy Cantrell).

enraged man near the very spot where Ella May was murdered was jarring. Only Ella May hadn't faced jut one hate-filled man with a gun, but dozens, and one of them shot her.

Due to our previous experience, I had some reservations about our next attempt to investigate Ella May's history. Based on information from Bobby Bolch, our group drove to the location where Ella May lived in Bessemer City. Ella May's home was destroyed decades ago, but we held hope that there would be remnants of her life there. A modern home with a well-manicured lawn now resided on the property. A friendly middle-aged woman answered the door. I explained to her that

I believed my great-grandmother's home once stood on her property. I asked her if she had a spring or stream on her property and was over-joyed when she said yes to both. I was relieved to find her extremely friendly and curious. She allowed us on her property, took us to the spring and stream, and explained what little she knew about the prop-erty. It felt surreal standing next to the water source Ella's family used, where a man once tried to poison the family. The property owner gave us contact information and graciously welcomed us back again. Her husband later talked to Kimberly. He told her there were indeed rem-nants of a historic house and that we were allowed back to explore.

We went back again. I was told by Bobby that Ella May's house was to the left of the current owner's home, but we were guided to the right. So I had some reservations about whether the pile of bricks located in some deep brush was indeed the remains of Ella's home, but clearly it was a home from that time. It was the only spot on the property that had any artifacts, so we investigated. The base of a fireplace was distin-guishable. Old bricks were strewn about. Mason jars, rusted tins, and upon further investigation a wash basin, were discovered. This artifact moved me the most. I could picture Ella May, or perhaps her eldest, Myrtle, using one like it. The springs to two beds were discovered as well.

After investigating the site, we explored an ancient lithium plant that was adjacent to the property. The plant existed prior to 1929, so surely it was a place the family explored. Near it were some small cas-cading falls that spilled into a little peaceful pool. Somebody in the group made the comment that Ella's children would have come here to swim.

The Ella May Wiggins Memorial Committee is now a nonprofit organization. Facebook connections on the Ella May Wiggins Memorial page have grown to nearly 1,000 individuals. A website has been estab-lished to educate the public at ellamaywiggins.org. Steve Pi has been selected by the committee to create a life-sized bronze statue of Ella May. Currently the statue is under construction and the committee is pursuing funds.

I am grateful to the multitude of sources I used to compile Ella May's story and the story of the Loray Mill Strike. I gained a wealth of information about the strike from books, especially from John Salmond's *Gastonia 1929*. The most in-depth information I learned about Ella May's life was from Lynn Haessly's unpublished thesis, "Mill Mother's Lament." Haessly interviewed my grandmother Millie shortly before her death, and as a result, many of Ella May's children's experiences are found in her thesis. I reviewed every page of every paper from the *Gastonia Gazette* for over a year to get additional insight. It was an awakening experience to see the actual inflammatory words that were printed as the events unfolded. Many family members retold their small recollections of family history.

Margaret Larkin helped keep Ella May's music and spirit alive. Larkin was a journalist who followed the late-twentieth-century strike scene. She started writing about Ella May and her music after witnessing her perform in Gaston County at a 1929 union gathering. After the performance, Larkin actively pursued the retrieval and documentation of Ella's ballads, later publishing them. Larkin too later performed some of Ella's songs in New York City. The two outside leaders in the forefront of the strike, Beal and Buch, both wrote extensively about their experiences. I found their perspectives valuable, especially Vera's, for no one published was closer to Ella than she.

I want to thank those who aided in my journey of discovery. First and foremost I want to thank my mother. She has not always understood me, but has always been proud of me, and has been willing to share family stories. I am thankful to dear friends such as Christine Bowman, Roxanne Newton, Allen Millican, and Tammy Cantrell, who have helped me gather research. I am grateful to Nick Ippoliti and Tripp Howell for reading and giving me feedback on early rough drafts.

This book would not have been possible had it not been for those who believed in me and my project. My former boss and friend Julie Waltman had no doubt in my ability to complete such an endeavor and encouraged me to work on it. Years later, George Loveland's own admiration for Ella May reawakened my passion for this project.

Generous resources have been shared from the *Gaston Gazette*, Allen Millican from Millican Pictorial History Museum, Tammy Cantrell from Encaptured Photography, of course my family, and many others. Most importantly I am forever grateful to the life of Ella May and others like her. If it weren't for Ella May and other brave activists like her, we wouldn't enjoy the quality of life that we have today.

Prologue:
The Legend

Early in 1929, the communist-led National Textile Workers Union (NTWU) focused on organizing textile workers in the South. They searched for the ideal location to start their efforts. Leaders soon found themselves at the six-story Loray Mill in Gastonia, North Carolina, the largest textile mill in the world. This became the staging ground for one of the most heated, heart-wrenching labor disputes of the time. Before it was over, beloved Police Chief Orville Aderholt and symbol of striking mothers, balladeer Ella May, would be killed.

Twenty-eight-year-old Ella May emanated strength and an air of dignity. She demonstrated native intelligence and raw talent. She picketed on the line, sang at union gatherings, and at times would speak to the masses. Today most who are familiar with her story see her as an inspirational woman who spoke up against the injustices of society. She is portrayed as one who, despite personal hardship and heartbreak, rose to confront a powerful, established order. She became famous for moving audiences with her music, the raw personal ballads and assertive songs calling for change. She is now grouped with the likes of Aunt Molly Jackson and Woody Guthrie, celebrated folk musicians and pioneers of social justice.

By contrast, during the post–Civil War South, in the pre–Depression era, within her own community she was considered a woman of ill repute. She spoke her mind against authority; she was brash, blunt, and did not follow the socially accepted behavior of a woman. She was associated with rabble-rousing and threatening the very livelihood of her neighbors. She intermingled with those outside of her own race. This young unconventional woman also associated with those who were considered a serious threat to American society itself: communists.

Loray Mill, ca. 1940 (courtesy Millican Pictorial History Museum).

She was and is to her family, my family, simply a mom, grandmother, or great-grandmother. She was a tiny mountain woman who had no formal education, the poorest of the poor, trying to keep her hungry brood alive. She tried to survive the best way she knew how while clinging to the hope of a better future for her children. She wasn't an untouchable, an ideologue, or a dangerous threat, resented at a distance. She was real, both flawed and respected, and most importantly dearly loved.

Ella May's impression on others was far-reaching and deep. Folk singer Woody Guthrie explained, "She never had much schoolin', but she knowed people. That's a heap better than a knowin' books." Professor and writer of history Jacquelyn Dowd Hall stated that Ella May was "the most well-known protagonist in the 1929 strikes." Close friend and strike organizer Vera Buch Weisbord wrote, "Ella May Wiggins has become a legendary figure in the history of labor struggles and her songs have found a place in the legacy of fighting textile songs." Newspapers such as the *New York World* raised her status as high as "the Joan of Carolina."[1]

Ella was a pioneer in organizing black workers. She is a legend in textile, union rights, feminist, and Southern history. She resonated and resonates with struggling workers. Ella May became a magnetic figure to a vast number of people. That, coupled with the combustible historic circumstances in which she found herself during the Loray Mill Strike that led to her early demise, makes her a legendary figure in American history.

The Loray Mill Strike has its own special place in history. It was the largest communist-led strike on American soil. It infiltrated Gaston County, the heart of the textile industry, the industry responsible for rebuilding the South. The strike branched out from Gastonia, to other neighboring Southern textile communities, such as the town where Ella May lived, Bessemer City. The communist union's presence was not welcomed by the majority. One hundred or so community-minded individuals bonded together and formed a vigilante group, the Committee of 100, to defend Americanism. This group, spurred on by the rhetoric spouted in the local newspaper, the *Gastonia Gazette*, terrorized strikers and their leaders. The breadth of this terror was reached the night of Police Chief Aderholt's murder mistrial, when mobs of hundreds combed the region indiscriminately beating strikers and destroying their property. The extremity of the terror was reached the day Ella May was shot and killed.

The Loray Mill Strike of 1929 was well documented. Its controversial nature made that inevitable. The touchy issues faced by those involved led to the writing of much opinionated material. Union and women's rights leaders who wrote about the strike saw native textile workers as social rights activists. They wrote about how workers were fighting for a social cause. They implied that workers' idealism and concern for social ills made them want to work towards radical change. Mill owners and much of the community saw strikers as lazy, wanting to enjoy freedom from work. This view is well documented in the *Gastonia Gazette* and textile literature of the time. They also believed that the cunning NTWU warped ignorant workers' minds into believing the ills of communism.

I tried to the best of my ability to weed through the various slants in order to express the truth as accurately as possible without adding too much of my own personal bias. However, the nature of my relationship to Ella May makes it impossible for me to be completely impartial. So this book represents the life of Ella May and the strike that took place in 1929 as accurately and as objectively as possible through the eyes of one of her great-grandchildren. I hope this story told from the perspective of one of Ella May's descendants makes the historical events more tangible to the reader, and more applicable to his or her own family's experiences.

History has everything to do with perspective. The past is not limited to one perspective or belief, but many. By being open to multiple perspectives we come closer to finding the whole truth. Everyone participating in the events of the Loray Mill Strike was American in his or her own way. Mill management, outraged community members, strikers, and even communist union leaders were American citizens participating in their own personal pursuit of the American dream. The United States has historically symbolized hope, the possibility of a better life for all, regardless of social and economic status.

Both individuals and whole nations in the pursuit of betterment find themselves in periods of time of profound change, confrontation, and crisis. The Southern Piedmont in 1929 was at such a time in history. Profound change was forced upon the South following the Civil War. The loss of slave labor launched the development of the textile industry in the region. Agriculture could no longer sustain the economy. Massive groups of farmers from the Piedmont Region and the Mountain Region gathered together in large mill communities to work long days in the textile mills. The textile industry depended on global markets. Demand for cotton workers dropped after World War I as advancing technology was reducing the labor needed. Confrontation and crisis developed when expectations of the mill owners and the local communities' return on mill investments remained high while mill workers' expectations of economic improvement remained. Both could not be sustained. Pressure from both sides eventually developed into violence.

The children and grandchildren of those who experienced this turbulent time were unaware of what their families had gone through. Too painful, too bitter were the memories held by their parents and grandparents to share with the younger generations. Many children and grandchildren, when sensing the anguish of their parents and grandparents when they did ask questions, stopped asking. The generations farthest removed from the situation, less exposed to the pain, are the most eager, and better equipped to uncover the past. Time has elapsed, and now not just historians, but the grandchildren, and the great-grandchildren, the great-great-grandchildren of those involved, the Piedmont Community is open to learn from the uncomfortable past.

History in its barest traditional form, in the sense of dates, names, and events, has had very little meaning to me, and I believe I can justly say so for most individuals. History is lifeless without a framework, a setting, an experience to tie you to the information. While attempting to understand Ella's life, her daily toil, I was pulled into the history of labor, industry, culture, gender, racial equality, and government. As I wrote I jumped back and forth between the direct events occurring in Ella May's life and the chronological story of the Loray Mill Strike. Background information regarding events, people and issues was inserted in order to create a better understanding. The scope and depth of historical events affecting Ella May's life and the strike she was involved in are impressive.

Violence erupted across the South in a number of textile towns in 1929. The 1929 strikes and resulting trials were referred to as "one of the three major labor events in the twentieth century." Strikes elsewhere were organized with no overt communist ties. Some, such as in Marion, North Carolina, where six strikers were killed, were bloodier than the Gastonia strike. None, however, obtained as much attention and caused as much controversy as the Loray Mill Strike. The Loray Mill Strike was seen either as a communist invasion or as a symbol of capitalistic wrongs in America. News of it reached Russia, England, Denmark, Germany, and other countries across the world. Protests were staged in many of them as the 1929 events unfolded. The Loray Mill Strike was a legend in its time.[2]

1

The Beginning

Ella May's name has been a source of confusion. Ellen May, Ellie May, Ella Wiggins, Ella May Wiggins, Ella Mae Wiggins, Ella Mae May, Ella Maise are all names Ella May has been identified by. Ella's current gravestone, erected in 1979, is incorrectly inscribed Ella Mae Wiggins. Mae was not her middle name. May was Ella's maiden name, Wiggins her married name. Ella referred to herself as Ella May for the last few years of her life. That is the name her co-workers, friends, and neighbors knew her by. The signature on her union ILD card from Bessemer City, North Carolina, is signed that way. Strike leader Fred Beal said that he was surprised to find that Ella's last name was Wiggins: "We had always known her as Ella May." The *Gastonia Gazette* also printed in Ella's funeral notice that "Mrs. Wiggins [was] better known as Ella May." Ella's independent nature and disassociation with her husband led her to revert to her maiden name. Due to Ella's rejection of the name Wiggins throughout this book, Ella is not referred to as Ella May Wiggins, as in the majority of printed material about her, but rather Ella May, or Ella. Ella May's oldest son Clyde too reverted to his mother's maiden name. The rest of Ella's children kept the name Wiggins.[1]

Ella May grew up in the remote Smoky Mountains, part of the Appalachian Mountain chain traversing the eastern United States. The southern highlands that Ella called home during her youth were majestic and mystical, changing color depending upon the time of day, weather, and season, a misty haze often surrounding them. The glaciers that once raked the northern mountains never reached the southern Appalachians. As a result the soil in the southern mountains was more rich and hospitable. The southern mountains' foliage was more dense and varied than the northern part of the Appalachian chain. Crops flourished high in the mountains.[2]

Streams, cascading waterfalls, and springs were plentiful there too.

It was hard to travel anywhere in the mountains without encountering some type of running water. The water was convenient for drinking and building spring houses, but flooding and drowning were constant concerns. Young children especially had to be watched around the open water. Young Ella loved to swim and had the habit of jumping into water when she got the chance, much to the chagrin of her parents. She would convince her less eager siblings to join her so they wouldn't tell on her. For those not looking to get wet, simple bridges made transportation possible. They could be fallen logs or several planks of wood pulled together. On rare occasions a handrail could be found. Bridges were not professionally constructed, so they were not especially sturdy or safe.[3]

Both early European settlers and the last of the Cherokee nation had sought the mountains as their retreat. Immigrants originally arrived in waves in Pennsylvania such as the Scotch-Irish in 1682 and the Germans in 1740. These hardy adventurers traveled southward down the Appalachian Mountains, making it as far as Tennessee. These mountain people who isolated themselves in the Smokies had their own distinct character. The independent, hard-working men and women were descendants of pioneers such as Daniel Boone and Davy Crockett. They clung proudly to the mountaineering ways of their ancestors.[4]

Life was tough. Everything that was needed was created on site on individual farms. Food, clothing, and furniture were all products of back-breaking work done by one's own family. There was no store close by where one could buy things. Little to no cash was used. A farmer's two hands, knowledge passed down by ancestors, and his own native intellect were what he relied on to survive. Self-sufficiency was a necessity in their rugged remoteness. Homesteaders were skilled farmers, hunters, carpenters, masons, and plumbers. There was pride in this culture of simple independent living, living as one's forefathers had. What was good enough for their ancestors was viewed as being good enough for them.[5]

Mountain folks were very much separated from the world. It was a dangerous, long, and difficult journey to go up or down the high rugged peaks with their unpredictable terrain and thick foliage. Rare outsiders entering into their world, trying to modernize mountain men and women, were viewed with suspicion. The mountain folk felt looked down upon by these uncommon opinionated visitors from the outside.

Charity was often confusing and unacceptable to those in the mountains. A stranger giving something meant something needed to be given back in return. Many visitors to the mountains felt perplexed after trying to be charitable. They were often refused or insisted on being given something in return, like a chicken.[6]

Roads were rare, and when they were worn into the dense wild land, they were rough, nearly impassable. Many living high in the remote mountains never saw an automobile, train, or airplane. Families traveled by horse and most often by foot, and more often on narrow winding footpaths rather than wider roads. The main purpose for traveling them was for visiting. Mountain folk would travel great distances to visit their neighbors or kin. Small, roughly built one- or two-room log cabins were always open to guests, expected or not. Privacy was not highly valued. Visiting was the main source of recreation. What little there was to eat was always graciously shared with others. Gatherings always ended in the evenings around the fire accompanied by folk music. Guitars, banjos, or fiddles would play common melodies as family members and guests would sing well-known ballads. Music was entrenched in the mountains. It allowed an outlet of expression to those so closed off to outside communication. Not only during larger gatherings, but in evenings with one's own family, or during daily work and toil, mountain music would rise up. Mountain folks would sing centuries-old ballads, as well as mountain melodies their neighbors or kin had created, or would reconstruct preexisting songs into something unique to fit them. Their own experiences and feelings were communicated in this way.[7]

Despite the isolation and lack of communication with the outside world, or more likely because of it, men and women in the mountains felt very free. They felt free to build homes and grow crops where they liked. They had no restriction on where to hunt, fish, or cut down trees, or limit as to what numbers they could take. The use of these resources was seen as a man's inalienable right. They had considerable work to do, but they chose what to do, when, and how to do it. Men and women in the Smoky Mountains were strong individualists. Equally important to being independent was the environment itself. Where they lived and worked inspired a feeling of freedom. The mountains were picturesque, the land diverse and nurturing. The air was pure and the skies clear. Working under these conditions stirred the soul. Life in towns, with buildings pushed close together, with greenery ripped away, and garbage and sewage spewing down the streets, was not even remotely as attractive. Work inside stuffy cramped buildings was not nearly as desirable as work out in the open air. Many mountain men and women, when they did leave the mountains, felt trapped in by urban life.[8]

Ella May was born within the mountains along the Tennessee and North Carolina border in Sevierville, Tennessee, on September 17, 1900. Her sixteen-year-old mother Catherine Maples May's heritage was Cherokee and her thirty-year-old father James May was Dutch. She was born into a blended family that included a half-brother, half-sister, stepbrother, and stepsister. The next May child, Samuel Lee Wesley (who was known as Wesley), was born six years after Ella in 1906. Two more were born following Wesley, one in 1908 and another in 1915, and both died during childhood. The first of those siblings, closer in age to Wesley, was a brother named Lewis (Lum) Jackson, whom Wesley accidentally shot in 1914 while the two were playing cowboys and Indians with a pistol they found in the woods.[9]

Eight-year-old Wesley and five-year-old Lum mistakenly believed the gun they discovered was empty. One shell remained. The boys took turns with the gun shooting at each other. After Lum had a turn he gave the gun to his brother. When Wesley aimed the gun at his brother and pulled the trigger the pistol fired and Lum collapsed to the ground. Lum lived for four painful days with his injury before dying. The family buried him themselves. Medical care was hard to come by in the mountains. There is no telling how long young Lum had to wait for treatment, what type of treatment he received, or if faster and more advanced methods could have saved him. Often times medical ailments were treated by mountain medicine, ancient herbal cures passed down by Native American or early European ancestors.[10]

While more modern health care may not have saved Lum, it surely would have saved his younger sister. The youngest May child to pass, four-year-old Hattie, died from uncinariasis, or hookworm disease. The doctor who recorded her death noted that the whole family was infected. Medical records indicate that years later Ella's five-year-old daughter Millie too tested positive for this ailment. It is very likely all of Ella's children did. Children running around outside with bare feet like the May children were at especially high risk of infestation.[11]

The May family eked out an existence on their small mountain farm. The family lived on corn, yams, cabbage, beans, and apples, which they farmed and harvested. Game was plentiful, and James May would hunt rabbits and opossums. Wesley recalled his father being "a hard worker that provided for his family." Hogs were also raised by the family and slaughtered in the fall. Food was dried or smoked for the winter. All the children had to work on the farm in order for the family to get by. What few items the family couldn't grow or construct themselves, such as tools, shoes, and lamp oil, were traded for.[12]

The railroad entered the Smoky Mountain area at the end of the 19th century and forever changed life for the Mays and most mountain families. It created a reliable network of transportation in and out of the mountains that enabled the development of the logging industry. Commercial logging was established there by the early 1900s and Northern capitalists bought vast tracts of land for their businesses. Logging employed thousands of workers and became the center of the Southern Appalachian economy.[13]

Before commercial logging entered the mountains, farmers used their trees as cash crops. They would cut down logs in the spring and float them down rivers to small local mills. After the logging industry entered the mountains, the widely practiced and entrenched subsistence means of living in the mountains became nearly extinct. Families across the area gave up their traditional means of living on farms for the wages found in industry.[14]

Men residing in the mountains were well suited for commercial logging, having intimate knowledge of the difficult, steep, thick terrain, and having experience cutting down the hardwoods found there. In 1910 the May family joined the new booming industry. Thousands of men across the country also came to the once-isolated mountains for work. At this point ten different lumber companies were canvassing the mountains looking for timber. The May family worked for the largest of these, Champion Fiber Company, which spanned both the Tennessee and North Carolina sides of the mountains.[15]

Once mountain folk such as the Mays were exposed to wages and industry, they became dependent upon them. No longer independent and free, no longer relying on their own two hands and wit, mountain families counted on the dollar for their meals, their clothing, and their shelter. With this monetary dependency, pride in craftsmanship was brushed aside in favor of ease and convenience. The mentality "what was good enough for my forefathers is good enough for me" was abandoned. Items desired rather than needed, such as toys, jewelry, fancy clothes, and linens were bought.

While Ella's father James May was employed as a logger, the family lived a transient lifestyle, moving from camp to camp to follow where the work was. They spent some time in the Nantahala Gorge, in Swain County, North Carolina, later moving to the camps around Andrews, in Cherokee County, North Carolina. James logged while the women in the family took in laundry for the single male workers. The women washed the clothes by using lye soap and stirring the dingy items in a big black kettle hanging over an open fire. Ella's younger brother Wesley also made a wage through the logging company. He carried water to the working men, learned the logging trade through observation, and eventually earned the right to help his father cut down trees.[16]

Both the masses of single men and the bands of families that followed their husbands' work received lodging from the lumber companies. Either occasional old mountain cabins bought with the land, or countless boxcars fixed up like homes, were the temporary housing. The boxcars were made out of rough lumber. Tar paper was used as insulation and linoleum covered the floors. Most were painted barn red on the outside. Boxcars such as these, set up with modern conveniences such as coal-burning heaters and curtains, were often preferred. The boxcars, which could only be moved by railroad flatcars, could be no longer than fourteen feet and no wider than eight feet. Clusters of these red boxcar homes littered the side of the railroad tracks, clinging to mountains, creating temporary logging communities.[17]

The makeshift logging campgrounds were where Ella first publicly displayed her talent for singing. She was confident and outgoing, and loved being the center of attention. Ella had chestnut hair and hazel eyes, was considered a pretty girl, and was well-liked by the men in the camp. In the evenings outside by the light of a fire, she strummed her guitar and sang popular ballads for the men who were worn from work. She sang "Little Mary Phagan," "Lord Lovel," and "Fair Margaret and Sweet William," and others.[18]

One logger who took a special interest in young Ella was John Wiggins. John's family, like Ella's, left life on a family farm to explore life in the logging industry. He was originally from Andrews, North Carolina. He was seven years older than Ella and a single father of a three-year-old named Lula, a daughter from a previous marriage. Lula's mother had died from a common danger of the time for young women, childbirth. John was a ladies' man, a smooth talker with penetrating eyes, and seventeen-year-old Ella was instantly enchanted by him. Ella married John around 1917 and soon after gave birth to a daughter, Myrtle, in 1918 at a logging camp in Swain County, North Carolina.[19]

A year after Ella's first child's birth her forty-nine-year-old father died in a tragic logging accident. James May was cutting down a log on the side of a mountain for the Champion Fiber Company near Andrews and misjudged the direction of the log's fall. Thirteen-year-old Wesley was by his father's side at the time of the accident. Wesley had foreseen the danger,

Men farming in Andrews, North Carolina, likely between 1907 and 1916. The Wooten clan is on the left and the neighboring Wiggins clan is on the right. John Wiggins is second from the right (courtesy Kimberley Hallas, restored courtesy of Tammy Cantrell).

and warned his father that the log may fall his way, but James told his son not to worry. When the log came down it plummeted towards James and he was unable to escape. Wesley immediately raced down the mountain to get help, but his father had been killed instantly. The family was emotionally and financially devastated. Wesley did not earn enough to sustain a family, so Ella's mother and brother moved in with Ella and John and their two girls. The clan continued to follow work in the timber industry. Not long after her husband's death, Catherine May became chronically ill. Ella nursed her mother, but in 1920 Catherine's kidneys began failing. She lived in a coma for a month and a half before dying from her illness.[20]

The pay from lumber companies came at a great price for Ella and her family. Not only her father, but her husband John too was later involved in a logging accident. He slipped while stepping on a log and crushed his leg. His injury was less serious, but still disabled him from working. He lived the rest of his days with a limp. There was little reason for Ella and her family to stick with the rambling logging communities after her husband could no longer work.[21]

Ella gave birth to her first son Clyde in 1921 in Cherokee, North Carolina. Not long afterward, Ella left the mountains with her family. Countless other families were leaving around this time as well. The need for loggers was dwindling. Families had to find other means to survive. The logging industry never intended to stay indefinitely. It was concerned with gaining maximum profits for the short run and not with long-term consequences. Once supply dwindled, the industry left as quickly as it had arrived. The logging companies did not reforest the land, and much of the mountainous region was devastated. Land erosion and other environmental problems soon followed.

Loggers had few options. They could return to their neglected farms or find new work. There were jobs in the coal mines of Virginia and textile workers were needed in the Piedmont. Most loggers chose to work outside the mountains rather than return to their abandoned farms because they held hope for an easier lifestyle. Farming was difficult prior to the railroad and logging infiltration. It became almost inconceivable after the destruction of the mountainside and after the mountain men and women acquired a taste for cash. The pull from the mountains was heavily influenced by financial reasons, but the sense of adventure was also alluring to young families such as the Wiggins family.[22]

Like thousands of others, the family accepted an offer from a mill scout to work in a textile mill. The scout gave the family money for traveling expenses. Ella chose the mills because unlike the mines, the mills hired women, especially young women. No group was affected by the proliferate textile industrialization of the South more than young women. Opportunities for women to earn income were scarce at this time. Women were expected to be the domestic center of the family. Exceptions included teachers and clerks for the middle class, and laborers in factories and house servants for lower class women.[23]

At a very young age Ella observed the two biggest social and economic transitions to occur in the mountains since the Europeans first arrived during Colonial times. Ella was originally part of a self-sufficient community consisting of hardy, industrious, independent pioneers who depended only on the land for subsistence. Then she observed the invasion of huge logging companies and witnessed the mountains being transformed. After that she participated in the great exodus of mountain people to the coal mines and textile mills.[24]

The timing of the departure was ideal for Ella, for she needed to support her family. She could no longer rely on John for subsistence. Finding employment as a cripple a difficult and depressing task for John. He took to the bottle for comfort and left his family for long periods of time. Ella said this was so he would not be a burden to the family. Testimony from family members indicate that his trips away seemed far more self-serving. Millie, Ella's daughter, said that she remembered her father being home just long enough to get her mother "in the family way." Millie said that on one occasion Ella asked John to get some groceries for the family. He left to get the groceries but then didn't come back for weeks. When he returned he was without food and broke. Ella never spoke poorly of her husband and portrayed him as a victim of circumstance. Mary May, who married Ella's brother Wesley, portrayed John in a less favorable light. She said, "He was the biggest liar around. He'd tell all kind of tales. He wouldn't try to provide for her and the children." John was in and out of the family's life from the time of his marriage to Ella in the mountains until his last child with her was born in 1926 in the mills. Ella eventually started relations with Charlie Shope. She never obtained a formal divorce. Divorces were an extreme rarity at the time.[25]

The family did not move directly from the mountains to the mills. Ella took her family first to a plantation located in Spartanburg, South Carolina. Here, at Millie Littlejohn's plantation, Ella had her first encounter with black workers. She worked side by side with them in the fields picking cotton. Unlike most whites born and bred in the Piedmont she had no prejudice towards them. This is likely due to her being born and raised in the mountains, where everyone living and working in proximity to one another is seen as an equal. There were no classes or races in the mountains. Everyone living in the wilderness had an equal footing. Racism or classism were not expressed or practiced.[26]

The work Ella experienced on the plantation was hard, but she didn't mind, for it was also at the plantation that she had another first experience, that of being the primary breadwinner. Ella was proud of her ability to support her family and not have to rely on another for subsistence. She admired the independent female owner of the plantation, Miss Millie, and named her child born there in April of 1923 after her. Millie was the third child born to Ella, the first outside of the mountains.[27]

2

The Mills

Ella eventually made it to the textile mills and worked as a spinner to support her ever-increasing family. The family continued living a transient lifestyle. Ella started off in the textile town of Cowpens, South Carolina, working for the Cowpens Manufacturing Company. In Cowpens, Ella gave birth to two children, possibly twins. She had a boy and a girl who both died soon after being born. After leaving Cowpens, Ella and her family moved to the center of textile manufacturing, Gaston County. During this time Ella lost another child, most likely the child Millie remembers as Thelma. They lived for a short period of time in Lowell, where a baby boy named Guy was born in January of 1925. He died from pellagra about a year and a half later while the family lived at the Rex Village in Gastonia. Son Albert was born at the Rex Village in June of 1926 shortly before Guy's death. The final child born to Ella, Etta Charlotte, was born in Bessemer City in November 1927. Bessemer City was the last town the family resided in.[1]

The last six years of Ella's life were spent in the four communities mentioned, perhaps more. Moving from town to town, often on foot, was physically demanding and dangerous for young malnutritioned children such as the May children. The decision to move could not have been made lightly. Despite the risk, Ella repeatedly left a textile job in one town in search of a textile job in another. Her situation wasn't unique. Turnover was high in the textile industry, in some mills averaging as high as 20 to 40 percent of the workers. High turnover was attributed to a combination of factors. Some were self-imposed, such as choosing to leave due to discontentment over conditions, pay, or treatment. At other times the decision to leave, especially for young mill mothers, was not of their choosing. Women would be fired for taking time off for childbirth or for taking time off to take care of sick children. In Ella's case there was yet another reason for being fired, and that was for brashly speaking her mind, which she had a reputation for. So Ella and her family became accustomed to roaming the Piedmont in search of employment.[2]

Mills depended upon the workforce, both the transient and fixed workers, and in turn so did the Southern communities in which they were built. Furthermore, the country as a whole depended upon the Southern textile workers' output. In 1920 as much as 80 percent of all the fine combed yarn produced in the United States was produced in Gaston County, the epicenter of the Southern textile industry. The first small textile mills began sprouting two years after Gaston County's formation. The Catawba and South Fork rivers supplied water power. The railroad entered Gaston County in 1872 and expanded the new textile industry, moving raw materials in and finished goods out of the area.[3]

Gaston County has a rich history despite its late formation. The county did not exist until 1846, when an act of the state legislature created it. Gaston and many other counties were carved out of existing counties as communities grew and their needs diversified. Gaston County emerged from Lincoln County, the South Fork River being a geological dividing line. Gaston County is named in tribute to Joseph Gaston, a former state Supreme Court justice and lyricist of "The Old North State Song."[4]

The textile industry was not the first industry to make its mark on the county. Initially Gaston County was known as North Carolina's center for whiskey distilling. During the county's infancy, distillers were widespread and a common advertisement phrase was "all you can drink for a nickel." By 1888 the temperance movement had significantly influenced the decline of the whiskey industry. Slowly capital was taken out of distillation and moved toward other industries, such as cotton textile manufacturing. In 1903, when liquor distilling became illegal, the textile industry's growth exploded.[5]

Like the mountains and much of the South, Gaston County's heritage was Scotch-Irish and German, descendants who originated from Pennsylvania. The native inhabitants, Cherokee and Catawba people, had long been displaced or eradicated. Most Cherokee had retreated west towards the mountains as white settlers encroached on their homeland. Catawba tribes disappeared as a result of exposure to Caucasian illnesses, mainly smallpox.[6]

By 1925 Ella May was living in Gaston County. The county in which Ella lived for the remainder of her life had become one of the most industrialized of the South. Gaston County was coined the "Combed Yarn Center of the South" and referred to as the "City of Spindles." The county was the leading textile county in the region, and third in the nation. Most mills manufactured combed yarn, but other cotton products such as tire fabrics, ginghams, flannellings, sheetings, hosiery, and labels were also produced.[7]

Prior to the Civil War there was very little industry in the South. The Southern economy was almost strictly agricultural. The invention of the cotton gin in 1793 by Eli Whitney increased the efficiency of cleaning cotton incredibly, eliminating the work of fifty workers per machine. At the same time, the industrialization of England initiated increased cotton demand. Both events made cotton production a lucrative agriculture industry in the fertile South. It also sealed the area's dependence on labor, specifically slavery. The institution of slavery made life for the free laborer very difficult. Unable to compete with free labor, the poor white farmers were pushed to less fertile land and into the mountains.[8]

When slavery was abolished following the Civil War, the white farmers reemerged and the plentiful source of black labor was shunned in the industry of manufacturing in preference of a white workforce. Blacks struggled to find their place in this new economy. Newly freed black men especially had a hard time making a living. Promises of employment made by Northern white men during the Civil War were found to be empty promises. Southern white men refused black men employment in favor of whites. Black women, while viewed as employable, had limited choices, mainly as domestic help.[9]

A new hope did emerge for white families. After the Civil War the financially devastated South's agricultural based economy could no longer be sustained. It could not survive without a plentiful free workforce. The South craved economic independence, so it had no choice but to change. Cotton production, which was once the center of Southern prosperity, was replaced by manufacturing based on the raw product, the textile industry. Manufacturing tied to the abundant and easily accessible cotton made sense. Textile mills soon symbolized growth and strength within the communities in which they operated.[10]

The period of reconstruction and industrialization of the South from about 1865 to 1885 was a difficult time. The wealthy forefathers of the "New South" following the "Cotton Mill Campaign" worked and sacrificed to develop their postwar broken communities into modern industrial centers. Mills became a source of pride and accomplishment, and every Southern town wanted one. Initially most mills were financed by the communities in which they ran. Stock was made available to locals in small incremental shares, which resulted in the common citizen's ability to own a piece of the mill. The editor of the *Gaston Gazette* in 1929 boasted: "Gaston County is the center of the fine combed yarn manufacturing industry of America. This industry has been built up through the toils and labors and thought and planning of men native to the soil who put their very lifeblood into this accomplishment. Thousands of Gastonians and Gaston County people, many of them workers in mills and on farms and in stores, have put their hard-earned savings into these very mills."[11] The textile workers who filled the mills had backgrounds similar to Ella's. Some, like her, had been recruited from the mountains and others had been local farmers. Both groups had experience working on independent farms.[12]

Independent farms became decimated during the post–Civil War era. Countless financially unsuccessful farmers provided the cheap labor the mills utilized. By the beginning of the 1900s Southern mills were able to pay workers 30 to 50 percent below comparable Northern workers. Hundreds of mills soon dotted the South. Southern mills underselling Northern competitors led to Northern capital too being invested into Southern mills.[13]

Investors found Southern mills to be a safe investment. There was an increased demand for cotton products during World War I from the military, and the industry profited. New mills sprouted in the South and especially in the heart of the Piedmont. By 1924 there were more mills in Gaston County than in any other county in the United States. The industry inadvertently was overextending itself, which eventually contributed to its downfall. After World War I, demand for cotton dropped and wages were cut, while workers' expectations heightened.[14]

Traditionally, Southern mill owners involved themselves personally in the lives and business of the workers they employed. Mill operatives were proud of their ability to provide for their native white workforce. Massive amounts of mill housing, consisting of small wooden cottages

Loray Mill housing (courtesy Millican Pictorial History Museum).

stacked upon brick piers, were constructed in grid pattern around the mills. Churches and schools were built, doctors and welfare workers were hired, and stores were operated, all upon the orders of mill owners. Even community buildings, athletic fields, and playgrounds were built, and social activities were organized under the mill owners' supervision.

Originally mill owners felt like they had a paternal relationship with the factory workers they employed. Moral behavior of the textile workers was of deep concern to them. The mill communities that were conceived and built by mill owners were a source of pride for both the owners and the local town community members who financed them. Drinking and promiscuous sexual behavior were discouraged. Mill owners prided themselves in making their communities both financial and moral successes. A worker could very easily lose his or her job due to behavior frowned down upon by mill management. Inside and outside of the mills, textile workers were watched.[15]

Mill villagers resented the control the owners tried to enforce on their personal lives. Many, having worked on independent farms, were driven by an independent spirit. They were accustomed to making their own decisions, rather than having most aspects of their lives decided for them. Whenever they were able to do so, mill workers took their own path. The churches were built by mill management and run by mill-appointed ministers. Often a textile worker needed a good word from one of these ministers to get a job. Mill workers would listen to ministers' rhetoric, but often did not agree with it or follow it. Religion was important to factory workers, but it and many other aspects of their lives were distinctly their own.[16]

Less auspicious habits frequented the lives of textile workers. Many working men and women were unwilling to give up their vices that enabled escape. Whiskey drinking was engrained in the mill community culture. Prostitution, while not as common, also remained visible around many mill towns. Workers risked their jobs while engaging in these activities. Anyone on a boss' payroll would have incentive to out a rebellious worker.[17]

Those providing welfare services were especially seen as the eyes

The four major churches of the Loray Community as they appeared in 1929. Clockwise from upper left they are: Loray Wesleyan (name changed to Firestone Wesleyan around 1935), Loray Baptist, West Avenue Presbyterian (name changed from Loray Presbyterian in 1920), and West End Methodist (name changed to Covenant Methodist in 1955). None of these buildings exist today (from *Loray, The Mill with a Purpose, Where the Boss is Your Friend*, published by the Loray Division of the Manville-Jenckes Company, 1929. Published in *The Centennial History of Loray Baptist Church, 1905–2005: Lighting the Way for 100 Years* by Timothy Craig Ellis, Trenton Creative Enterprises, Spartanburg, South Carolina for the Centennial Committee of Loray Baptist Church, 2005, courtesy www.vintagegastonia.com).

and arms of the bosses and were often unwelcomed. Workers believed better medical care, sanitation, organization, and services were not as much priorities to these individuals as keeping a watch out for the bosses. In the 1920s a sociologist who studied Gaston County mill workers questioned mill bosses and their welfare workers regarding employees' sentiments towards social programs the mills supported: "As a rule the mill help does not greatly value or appreciate in either sense of the word the efforts for their entertainment or advancement. First, they do not respond with any feeling of gratitude because they argue that all the money being spent on them is due them anyway as wages and some of them would rather have it in their [pay] envelopes."[18]

Over time Southern mills were increasingly taken over by Northern owners. These Northern owners were concerned solely with making profit and not with the welfare and morality of the Southern community. Welfare services were seen as undue expenses. Existing programs were cut back and few new ones, if any, materialized. Due to the lack of appreciation expressed by mill workers and the additional desire to cut back on spending, most programs were abolished by the late 1920s.[19]

Northerners invested in Southern industry due to the plentiful, cheaper, unorganized labor found there. The Southern work system was absent of the Socialist ideals and revolutionary thinking that an immigrant workforce had brought to the North. Even more so, Southern labor did not trust union organizers and their radical talk. Southern working-class folks especially did not trust strange-sounding, well-dressed Northern liberal union leaders who came down from the North and who were so unlike themselves. The Southern workforce's aversion to unions gave owners the impression that they were docile. However, workers were leery of both the Yankee union leaders who invaded to push their agendas, as well as the Yankee mill owners who invaded and valued them solely for the money they could make from them. Workers did not welcome union organizers easily or accept Northern owners' new practices and procedures. Civil War resentment of Northerners that still lingered was fed, not fading, with the inundation of Northern organizers and owners. Both owners and organizers ironically thought they were improving workers lives, but were resented for their actions.[20]

Workers weren't content with their economic situation. When independent means of protest did not produce change, workers slowly opened up to outside Northern organizers. Due to Southern workers' frustrations and limited options, Northern union leaders eventually found an audience that would listen. Even after Northern union leaders had their foot in the door, they did not find their task of organizing easy. Those they organized did not instantly embrace all of their ideals and ideas. The values found in middle-class educated union leaders brought up in Northern cities did not always match the values found in poor textile workers brought up on independent Southern farms. Union leaders encountered much reluctance from the workers they tried to help. Inevitably the interference and pushing of one class and/or culture onto another caused dissent. The interaction of members within the same group is a more desirable situation for those being urged to change.[21]

Mill workers felt most comfortable with other mill workers, with whom they shared a deep bond. Interconnectedness was created by both shared experiences and shared space. The proximity between neighbors in a mill community created a tight family-like feeling between residents. In times of hunger, illness, and need of childcare, mill families shared what little they had with each other. The feeling of connectedness was universal and timeless. Workers living in a mill village in 1900 felt the same nurturing sentiments towards those in their community as those living in mill housing in 1950. Neighbors looked out for each other's children and a mischievous child found it hard to escape the watchful eyes of the community. Most men and women raised in a mill community have precious memories of growing up in a nurturing, encompassing family. This interconnectedness between mill families deepened the impact a textile worker's words could make on another.

That is why, when unions were able to get a footing such as in the Loray Mill Strike, native organizers such as Ella May were so critical in the organizing efforts. Ella was able to express the union's message in a way that her coworkers felt comfortable with, that they could relate to, that they could support. Revolutionary talk with foreign words in foreign voices did not move workers, but regret and hope for their hungry children—shared in the form of the emotional, familiar mountain melodies—did.

The core grievances, pay and working hours, were the same for both workers in the North and South, yet where at this time much of the North was successfully organized, the South was not. The Southern labor community had expressed their desire for change not by the successful organizing of the masses, but by the longstanding traditional method of high individual turnover. Workers let their feet do the talking. Individuals left undesirable conditions and went to find employment within the same town, in nearby towns, or as far as several states away. Time and time again families such as Ella's would move. However, little improvement was found in this manner. Ella said that she had "worked in many mills" and had "found them all the same." By 1929 Ella May realized what she and the other workers had done repeatedly—leave to find a better situation—was not working. They had left unfavorable situations only to find the same conditions, or worse. This is why Ella and the other workers joined the NTWU. It was not the easy way out for them. It was a last and desperate final option.[22]

The American Mill No. 2 in Bessemer City was in western Gaston County and was the last mill in which Ella worked. This mill, like many others, went through a series of owners and name changes over the years. Originally the American Mill was named Whetstone Cotton Mill in 1903 when it was built. The owners were S.J. Durham, J.H. Wilkins, and L.W. Buck. The founders floundered in making it a profitable venture and in 1909 it was sold and turned into Huss Manufacturing Company. In 1920 it was sold again to the McLean family and became the McLean Manufacturing Company. The Goldbergs, owners of the American Cotton Mills Incorporated, bought the mill in 1925 and named it the American Mill No. 2.[23]

It was unlike the vast majority of the mills in the South for approximately half of its employees were black. In Gaston County about 15 percent of the population was black, but very few worked in mills. The Bessemer City mill in which Ella worked was an exception because most whites would not work there. The mill was a waste mill, paying extremely low wages under poor conditions, even for mill standards of the time.[24]

Most blacks in the South at this point in history still worked on plantations as their ancestors had. Mill work, while harsh, was still considered white man's territory. The few jobs that employed black workers in the textile industry were the least desirable ones, usually involving heavy labor, or especially dirty work. Socially acceptable jobs for black workers included being sweepers, porters, truck drivers, and elevator hands. Most often white textile workers would only catch a glimpse of black workers. For example, black workers commonly rolled cotton bales into a mill's opening room, broke the bales, and then would quickly disappear again. Other times black workers would be in the backdrop sweeping and cleaning the mill. Often owners would threaten to replace white workers with black workers if they got out of line. This method of keeping white textile workers in line incited white workers' resentment towards black workers.[25]

Ella May was unusual for a white woman in that she worked, lived, socialized, and even organized with black folks. From 1927 to 1929 Ella worked twelve-hour shifts at the integrated American Mill No. 2, and it was no doubt one of Ella's last resorts for her starving family; $9 a week was extremely poor pay. The typical cost for a four-room mill house at this time alone was $8. Ella, however, was able to find a home that had especially cheap rent. It was located on the outskirts of Stumptown, a community consisting of shabbily built homes that were erected for black textile workers and their families. A white family such as Ella's living in housing reserved for blacks was quite unusual since blacks and whites were almost always segregated from each other. On occasion they would work with one another, like in the Bessemer mill in which Ella worked, but these cases were rare in themselves. For whites and blacks to live so closely together, in the same community, was almost unheard of. Most whites, even the poorest among them, looked down on blacks and shunned interaction with them. Blacks too had little to do with whites. They did not trust whites, for good historical reasons, and shied away from them regardless of how well intentioned they seemed at first glance.[26]

Ella was different from most white folks her black community had known and they opened up to her. She saw them as equals, neither to be looked down upon nor pitied. They respected her gritty determination to support her family by herself. Ella's children recalled that their mother enjoyed socializing with her neighbors on her day off. She told her children that it was important to be neighborly. She was kind and well grounded, and her neighbors found her pleasant to be around. Despite these qualities the key factor that enabled Ella to bond with her black neighbors was the fact she was poorer than almost all of them. They knew this woman could relate to their plight of poverty.[27]

It was hard to find a white family poorer than a black family at this

Ella May's Mill-American Mill No. 2 in Bessemer City, ca. late 1920s (courtesy Millican Pictorial History Museum).

time in the South, even at the lowest poverty level. While standards of living at this time were poor for white workers, it was far worse for black workers. While the predicament of white workers was beginning to make headlines nationally, black workers' lives were still being ignored by the media and politicians. Ella was one of the few white individuals who could truly identify with some of the conditions black families faced because she suffered alongside them.

Ella's house was situated in the woods, close to a spring which was the family's water source. The spring fed a rambling little stream. The dilapidated building in which the family lived was low-priced, yet an entire wall was missing from the rear of the home. A tarp was used for protection against the elements. Ella's family, like countless other Southern mill families, not only had to contend with the heat, cold, and rain, but with great numbers of insects. They would breed underneath mill homes where pools of water would collect. Mosquitoes and flies were constant irritants in these homes without screens.[28]

Besides discomfort, there was an additional cost to pay for cheap housing, and that was the long walk to the mill each day. Ella lived a few miles from the mill where she worked. The closer mill housing was more convenient, and somewhat economical, but still too expensive for Ella's paycheck. Daughter Millie said that when her mother was done with work she wouldn't walk home; she would run, no doubt eager to reunite with her young unattended children. The children were often left home alone. John Wiggins was rarely around once the family left the mountains. Once the family relocated to Bessemer City in 1925, they never saw him again. Ella was gone much of the time to earn a wage for her family. The youngsters survived the best they could. Myrtle, Ella's eldest, only seven when her father left for good, was the main caretaker of the family when Ella was away. Myrtle would feed, clothe, and tend to the younger children's needs. Lula, the children's stepsister, was four years older than Myrtle and would have been a great help. However, she was sent to live with relatives when John and Ella's marriage fell apart.[29]

Younger sister Millie would sometimes help with the cooking and had to stand on the tips of her toes to reach the stove. She remembered that she could not see the top and burned herself on several occasions. The stove was very important to the children for the wooden home they lived in was poorly constructed and there were constant gusts of air coming in from the outside. During the winter the children spent most of their time huddled around the stove. During the hot Southern summers the heat was stifling. Millie and Clyde would often seek comfort in the nearby creek and take turns pouring buckets of water over one another.[30]

Myrtle was constantly caring for hungry and often sick infants and toddlers. With limited knowledge and almost no medical or food supplies, she tried to keep her younger brothers and sisters alive. Some of the babies made it to childhood; others did not. One hot summer night in 1926 little seventeen-month-old Guy, or Gay as the children called him, was up crying all night. A then eight-year-old Myrtle sat and rocked him for hours and hours until he stopped crying and then placed the silent motionless little boy on the bed. Myrtle didn't realize it, but Guy had died. Millie, who was three at the time, remembered, "When mama came from work, Myrtle was there, pattin' the baby. It wouldn't come back to life, it was Gay, and he was dead."[31]

The same month Guy died, the eighth and final baby born to Ella and John Wiggins was born in Gastonia, North Carolina. John was listed as the father on Albert McFalls Wiggins's birth certificate and also as the informant on Guy's death certificate. Soon after these events John left the family for good. "Pappy done tuk off," said Millie. Ella May then no longer referred to herself as Ella May Wiggins, but used her maiden name, Ella May. Ella left her married life behind her as she left the Rex Mill in Gastonia and headed to Bessemer City.[32]

48

Not long after John's departure, another man entered Ella's life. Bessemer City textile worker Charlie Shope appeared one night in Stumptown after being injured in a fight. Millie recalled, "Mamma took care of him. After that it seemed like I seen a lot of him." Millie didn't remember him fondly, saying he was "a little bit of a mean character ... he was a macho man." According to Millie on one occasion Charlie drew a knife on Ella, who "went right after him." Despite Charlie's hot and at times violent temper, he did help Ella financially support her brood while they lived and worked in Bessemer City.[33]

When Millie was questioned about the paternity of the youngest, Charlotte, and of Ella's pregnancy at the time of her death, she had no doubt in her mind that the father of both was Charlie. "If another man had messed around with my mamma, he would have broke their neck," explained Millie. Many years later when Charlotte was a young girl, Charlie arrived unexpectedly at Wesley and Mary May's house and talked to Charlotte, calling her "my girl." Charlotte, irritated, denied it and said, "No, it's not, sittin' over there on that stump is my daddy," pointing at her caretaker and uncle, Wesley. Charlotte in her younger years rejected the idea that she was the illegitimate child of Charlie Shope. At the age of twenty-nine her birth certificate was changed to read that John Wiggins was her father rather than Charlie at the time of her birth in November 1927.[34]

Characterwise, John Wiggins was hardly an improvement, having been an alcoholic who abandoned his family, and having

Ella May poses with her lover Charlie Shope shortly before her murder (courtesy Gastonia Regional Chamber).

49

a reputation as a ladies' man. However, he was the father of the rest of Charlotte's surviving siblings. More importantly, Ella May, while not in a relationship with him, was still legally married to John Wiggins during Charlotte's conception and birth. There was much shame in being a child out of wedlock at that time, and even more so when the mother was already married to another man. This is most likely the cause of the name change on Charlotte's birth certificate.[35]

Life was always difficult for Ella and her family. The extent of this hardship manifested in the heart-wrenching deaths of her children. Four of her babies died shortly after birth. Little Guy was the oldest, being seventeen months old when he died. One passed away at two months, another at three months, and another at a year old. Some died from pellagra, a disease characterized by fatigue, diarrhea, scaly skin sores and dementia, followed by death when left untreated, as in Ella's children's case. The disease is a result of poor nutrition caused by the absence of niacin or vitamin B3 in the body. Niacin is found in meat, dairy products, and some vegetables. Those with healthy diets had little to fear. However, for those who were poor and had extremely limited diets, pellagra was a constant worry. Pellagra was frequently found in textile communities. The surviving children always had their hair cut short, sometimes shaved in order to address the sores and scabs on their scalps. Millie's daughter Darlene recalled that these ailments were due to a common disease resulting from an insufficient diet. The likely culprit was pellagra.[36]

Vera Buch encountered a man who suffered from pellagra during the Loray Strike. She said he had reddish-brown, thin, shiny skin, along with a mouth full of sores; he was weak and unable to work, which troubled him dearly. The degree of disability due to the disease varied considerably. Some lived with sores and no other symptoms, while others were completely disabled, and about 40 percent of those who suffered from the disease eventually died from it. Today the disease is virtually unheard of in developed nations. Yet, from the time when the first Pilgrims arrived until the first part of the twentieth century, it plagued the poor in the United States. Today the disease can still be found in developing nations, especially those that depend heavily on corn for their diet.[37]

All of Ella's children, both infants and children, suffered from malnutrition. Millie remembered, "We didn't have enough to eat," when recalling her early childhood. Any disease in Ella May's children's weakened state could have been deadly for them. Ella's surviving children were lucky to have survived. Millie's medical records indicate that she suffered from diphtheria in 1926, followed by measles, chicken pox, and whooping cough in 1927. A deficiency of milk in her diet was also recorded on her records as well as testing positive for hook-worm disease. Poor families dreaded a wide range of illnesses. The ailments Millie suffered from as a young child were common; so too were croup, pneumonia, mumps, and the flu. Vaccines were not given, and treatments rudimentary and often ineffective. A healthy diet and practicing good hygiene were the best defenses. Unfortunately these were often absent in working poor families.[38]

Ella's children, like other mill workers' children, lived on a limited diet of corn meal, grits, beans, field peas, fatback, and on very rare occasions, green vegetables. The children almost never had milk, meat, fruit, or fresh produce, and what few items the family did have to eat didn't last long. The children were always hungry, so their mother would make them "sugar titties." "Sugar titties" is what Ella called her concoction of sugar and baking soda mixed together and placed into a rag which the children gnawed on. Apparently this was a common strategy struggling mothers used to ease their children's hunger pangs.[39]

The children were left alone while their mother worked, sometimes days, but even more often nights. There were only two beds in the house and the smallest children would sleep on them while the older children slept on rafters. While alone during the day they would try their best to fend for themselves. Clyde would lead the others hunting and they would corner an occasional opossum and stone and club it to death. Myrtle would then cook the meal for the children. Once or twice Clyde coaxed a neighbor's chicken to their porch with breadcrumbs. The breadcrumbs led straight to a propped-up tin tub which the children had tied a string to. The plan was when the hungry hen had made it underneath the tub to pull the string and capture it. On at least one occasion the children were successful and Myrtle was able to prepare fried chicken. The children quickly ate the meal and hid the evidence before their mother returned. Ella had strictly forbidden her children from stealing. The children rationalized that because the chicken wandered onto their property they weren't really stealing. However, they were not confident that

Ella would agree with their logic. One time she discovered what they had done. She scolded them, but was at the same time proud of her cunning children.[40]

Ella's children were dressed in worn, often dirty, ripped clothing and did not wear shoes. It was not uncommon for younger poor children to go shoeless. It was, however, considered improper for an older child such as Myrtle to go barefoot. A black neighbor, seeing that Myrtle had no shoes, gave her a spare pair of sandals. Myrtle wore the sandals when her mother was at work and hid them upon her return. Myrtle feared discipline from her mother, for Ella was a proud woman and did not accept handouts. Despite not having much, Ella followed a moral code and held her head high. She raised her children to do the same. She was not a woman who looked for the easy way out. Ella told her children that her neighbors were as poor as they were, and that it would be improper to take things from them. Ella never did discover that Myrtle owned shoes. After Ella's death Myrtle overheard others share their disgust that the oldest of Ella's brood did not wear shoes. This was a sore point for Myrtle, even well into adulthood. She was angered by their comments for in her mind she owned a pair.[41]

Many outsiders didn't understand why Ella's children didn't attend school. Schooling was offered in mill communities and many mill children did attend. Ella said that she was unable to send her older children to school because they were needed to tend to the younger ones. Her children didn't attend church, either, because Ella said they needed shoes to do so, which they didn't have. The plight of improperly clothed children was a common one. It wasn't uncommon for youngsters living in mill housing to forgo school due to lack of clothing. There were also a number of them who went barefoot, even through the winter.[42]

Parents had their own difficulties. The conditions in the textile mills in which Ella and the other workers spent much of their time were shocking by today's standards. A hazy, filmy light illuminated the mills, due to the dirt, dust, and inadequate light sources provided. The machines for spinning and weaving were immense, resounding, and dark. Conditions inside were unsanitary in the extreme. The limited lavatories were often unkempt and indecent for proper use. Floors were dusty, soiled, and drenched with tobacco spit from workers. The mills would be stifling hot in summer months due to the Southern climate, and constantly damp year round due to continually running humidifiers. The conditions of the mills contributed to a variety of physical and mental ailments that plagued workers. Respiratory diseases such as bronchitis, pneumonia, and tuberculosis were common. Many textile workers suffered from brown lung disease, or byssinosis, resulting from repeated inhalation of cotton fibers due to poor ventilation. Brown lung disease develops after years of cotton dust builds in the lungs causing a brownish appearance of that organ and constriction of the airway.[43]

The final mill that Ella May worked in, American Mill No. 2, was one of six or seven mills in Bessemer City. This particular mill was struck against longer than any other in Bessemer City in 1929. The owners were the Goldbergs, an immigrant Jewish family originating from Latvia, who relocated to South Carolina before moving to North Carolina. They profited from the mills due to their system of taking over poorly run and financially failing ones, and turning them into tightly run money makers. These mills, when run efficiently and tightly, would make huge profits in early years, up to 75 percent return on capital, after which a more normal rate of return would ensue. The Goldbergs would use the wealth from a newly organized mill to buy another one. In this way a steady cash flow could be secured. The Goldbergs were seen by prominent local leaders as outsiders unfamiliar with the proper Southern methodologies of running a textile mill. The family was a target of bigotry and criticism during times of striking activity. The Reverend C.J. Black, the Bessemer City minister who officiated over Ella May's funeral, called the mill-owning family "Jews who don't know anything about running a mill." The reverend said the company's exceptionally low pay scale "catered to a low class of people, the kind we don't want in Bessemer City."[44]

Despite the combustibility found there, the heart of the strike was not in Bessemer City where Ella lived, but in nearby Gastonia. Gastonia was the centerpiece of Gaston County and in the center of that centerpiece stood the Loray Mill. The Loray Mill was immense, by far the largest textile mill in Gaston County. When the plant was installed it increased the total export of cotton goods for the county by about 40 percent. In February 1900 the mill was erected, but it did not start operation until 1902. The stock of the plant was worth $1,000,000 and was financed mainly by Northern money, but had some local investors. The mill was never an extraordinary success, for it was constructed to make fabric for the Chinese market. This was a profitable venture for a short period up until 1903 when a major order was being shipped to China. During the time the shipment was to be made, China ended trade with the United States due to the Boxer Rebellion.[45]

George A. Gray and John F. Love were the two founders of the mill and their names together formed the name Loray. Gray was the operator of the mill and Love was in charge of the financial and promotional ends of the business. John Love was the son of Robert Love, who was one of the founders of Gastonia's first cotton mill with the Gaston Cotton Manufacturing Company. Robert's grandson Spencer Love became the founder of Burlington Industries, the world's largest textile company in the 1930s. John Love himself was one of Gastonia's first bankers.[46]

Gray had a lifetime of experience with textile mills. His was a rags-to-riches story which subsequent mill supporters loved to share. At the age of eight, after his father died, Gray began working at a mill in Gaston County as a doffer boy for ten cents a day in order to help support his family. He worked his way up the chain of command, becoming the superintendent of numerous mills before moving to Gastonia. In Gastonia he began investing his own capital into mills and by the time of his death he had influence in nine of Gastonia's eleven mills. Gray's life idealized what a capitalistic society could offer and reiterated the value of hard work in pursuit of the industrial dream.[47]

The founders, however, failed to finance the Loray Mill successfully and lost control in 1904 when major investors took control and reorganized. Control of the mill changed a few times before the Jenckes Corporation, originating from Rhode Island, bought the mill in 1919. The Loray Mill was the first mill in the county to be both owned and operated by Northern outsiders. Once controlled by Jenckes, the mill moved from producing cloth to producing tire and cord fabric to supply the growing automobile industry. This move increased the need for spinners and reduced the need for weavers. There were approximately 90,000 spindles at this time.[48]

In 1923 Jenckes merged with another Rhode Island company to form the Manville-Jenckes Corporation. Manville-Jenckes then owned seven mills in Rhode Island and two in Gaston County. The Loray Mill was expanded at this time and became the largest textile mill in the world. The brick mill, fortress-like in appearance, covered the area of two city blocks and was six stories high. The Loray Mill alone held ⅕ of the spindles in a county coined the "Combed Yarn Center of the South." Due to the size of the Loray Mill and the breadth of its parent company, Loray Mill management became increasingly removed from the lives of its Southern workers as time progressed. By 1929 the Loray Mill employed 2,200 workers, and 8,000 men, women, and children occupied its mill village.[49]

3

A Primed Situation

Strikes swept across the Piedmont in the early months of 1929. In North Carolina there were strikes in Pineville, Forest City, Lexington, Draper, Danville, Marion, and Charlotte, as well as Bessemer City and Gastonia. Thirteen mills in South Carolina went out on strike that year too. Other neighboring states, such as Tennessee, had their own share of idle plants. Elizabethton, Tennessee, was one of the first places to strike in March. Here the nation's attention was caught by the shocking vigilante tactics used by townspeople to try to expel the union threat. Marion, North Carolina, also made headlines; its unrest was more violent than that of Gastonia, and lingered longer.[1]

Tensions grew throughout the 1920s as textile workers across the South became increasingly dissatisfied with working conditions and wages. An increasingly strict, prison-like atmosphere developed in the Southern mills. Hank clocks were invented to measure the amount of yarn spun. This enabled management to set quotas. Stretch-outs, as they were known in the South, or speed-ups, as they were more commonly known in the North, became the new working mill philosophy. The stretch-out was the practice of extending an employee's responsibilities. More machinery was expected to be covered without the benefit of increased pay or incentives. A weaver in a textile plant, for example, could be made responsible for tending forty looms where he had previously been responsible for twenty-five. The additional burden of work did not allow for periodic breaks which workers especially from the warmer South were accustomed to. Nothing outraged workers more than this new practice.[2]

Striking, while not historically commonplace in the South, was not a new practice for those fed up with conditions. Workers across the South had often threatened to strike, and at times they did. Sometimes there was union support, yet many other times workers had no organized system to fall back on. Workers walked out en masse spontaneously out of frustration and desperation throughout the early 1900s. Independent walkouts hit a climax in 1929. Up to 10,000 workers left their jobs and struck without representation. Independent and restless, individuals refused to tolerate conditions they deemed to be unjust.[3]

A personal testimony to a spontaneous walkout was retold by K.O. Byers, one of the Gastonia strikers. Byers, while working in the Loray Mill, protested changes made by management. Specifically Byers resisted doing his own oiling and the increase of daily cleanup from one time to three times a night. At first the mill refused to give in, but in this particular instance the changes in the plant were reversed. Byers and the twenty-seven others who had left with him were asked to come back to work.[4]

Some unions came, fueled by workers' discontent, but none were able to keep a strong hold in any mill. Textile strikes were held but soon dissipated. Lack of funds from the unions was blamed. Workers not quickly and fully financially supported returned to work. Families had to eat, so those holding potential paychecks drew them back. When striking numbers dwindled, unions withdrew. Unions blamed strikers' lack of patience and union interest on failing strikes. Regardless of where the blame lay, each time unions left before demands were met, workers lost additional faith in organized labor.

During the textile industry's infancy in the 1880s and 1890s, the Knights of Labor, followed by the National Union of Textile Workers, organized textile workers in the South. In 1919 the American Federation of Labor's United Textile Workers Union (UTW) members arrived in the Gastonia area and organized Loray Mill workers who were protesting recent wage cuts. A handful of workers were fired for joining the union and a subsequent strike was held as a result. The strike lasted several weeks and involved 750 employees. However, the UTW left, accomplishing very little and crushing strikers' hopes. Gastonia textile workers became further disenfranchised by unions.[5]

Workers relied on their own creative methods of protest in Gastonia in 1928. Outraged at the extent of demands placed on them, Loray workers drove out one of, if not the most demanding foremen they had ever endured. He was literally forced out of town by a mob of workers to much jubilation from the crowds on the streets. During this street demonstration, textile employees carried a man in a coffin down the main street of town. A man lay in the coffin, which was labeled "Superintendent So and So." Eight men carried the coffin, and periodically the man being carried would sit up and declare "Six can do it," and two pall bearers would fall away.[6]

The new and despised methods of production in the late 1920s were the results of an industry amidst a recession. By 1924 the textile industry's recession was firmly entrenched. Technological changes, overproduction, competition abroad, and a constricting market all contributed to the stress in the textile industry. Too many plants were running in a tightening marketplace. Shorter, fashionable skirts of the time, together with man-made fabrics such as rayon, reduced the market for textile goods. By 1928 the piece system was increasingly being enforced in the mills in order to keep labor costs down.[7]

The average adult textile worker's pay in North Carolina in 1929 was around $13 a week; day shift workers worked sixty-six-hour work weeks, and the night shift worked sixty. During this time 24 pounds of flour cost ninety-nine cents, a loaf of bread cost seven cents and men's work shirts cost around seventy-nine cents. Southern textile workers on average were making 44 percent less than their Northern counterparts, according to a U.S. Bureau of Labor report. There was a surplus of labor in the South, so wages would no longer climb, and in fact in many areas wages dropped significantly. For example, in one mill wages dropped from $20 a week to $13. As a result there was an increase in the crime rate, prostitution, and drinking in mill villages.[8]

Opportunities for women to earn a livable family wage were severely limited at this time. While textile mills gave poor women the same opportunity as men to work, the money earned was not equitable. Women as a whole worked more hours in the mills and took home significantly less money. Not only women, but children too worked in the mills, and for even less pay. They worked in large numbers in the early years, yet were still visible into the late twenties. Some youth worked as much as sixty hours a week for less than five dollars a week. This was

Opposite, top: **Girls in the Loray Mill run warping machines in 1908.** *Bottom:* **Young women such as this one in Cheryville, North Carolina, were frequently hired as spinners and weavers (both photographs by Lewis Hine, National Archives).**

possible because many children under the age of sixteen lied about their age. These underage workers working under false pretenses were used as spares and therefore paid only half as much as a normal hired hand. Parents sent their children to work out of necessity to make ends meet. Furthermore, most workers had a history of working on farms where every member of the family, regardless of age, worked. Management was well aware of children working within the factory walls, but weren't willing to give them up. Not only were they a form of cheaper labor, but they were used for jobs that took advantage of their small hands and small bodies. Their ability to maneuver in small spaces was very useful.

The National Textile Workers union (NTWU) was very much aware of the situation in the South. Their headquarters was located in the North, and almost all their members were Northerners as well. The origination of the NTWU was the result of a policy change by the American Communist Party. Previously a philosophy of "boring from within" was practiced regarding unionization in the United States. The party members' Marxist ideology and views were passed to existing American unions in a passive, subtle way by American Communists joining these unions. Communist party members did not advertise their presence, nor try to impart their philosophy, but rather tried to influence policies and practices they favored. Communism, the philosophy itself, wasn't so much being preached to others as the practices that Communism believed in.[9]

This all changed in 1928. The new philosophy was that of a separatist approach; "dual unionism" became the new catch phrase. The NTWU was the first of a string of newly created unions. Its birth was in September of 1928 following the 1928 Sixth World Congress gathering at the Kremlin. The Workers Party of the Soviet Union was ready for more aggressive tactics in America. Two key goals for the NTWU were close ties to the Soviet Union and organization of textile workers, specifically in the South. A Communist revolution was always the final objective of top party officials and the motivation behind all party decisions. Textile mills were targeted as one of the first to be infiltrated, then all of manufacturing, followed by shipping and food production. The revolution planned for the United States, while covert and manipulative, was to be peaceful and nonviolent. Capitalism was what was under attack. An equal voice for everyone was the goal. Social status, gender, and race were no longer to be barriers of one's voice being heard.[10]

Some wondered why the NTWU did not start organizing in the North, where organized labor was already established. More liberal views were accepted and strikes were more commonplace and tended to be more successful. In addition, the North was a more familiar place to organizers. There were reasons why the NTWU preferred the South. In New England there were many nationalities, religions, languages, and cultural backgrounds. Southerners, for the most part, had a similar dialect, the majority were Protestant, semi-educated, uncultured in the sophisticated sense, and were ignorant of Communist themes and doctrines. This unfettered cohesion amongst the people made Communist leaders believe that the South would be easier to unite, that a central thrust here would be easier. In addition strikes were too commonplace in the North to receive the type of attention that the NTWU sought.[11]

The NTWU's first secretary, Albert Weisbord, was on the National Executive Board of the Communist Party and the textile representative for the party. It was Weisbord's responsibility to come up with a plan for organizing the South. Weisbord, like the majority of the leaders of his party, was an American-born Northerner. He was Harvard educated. While never a factory worker himself, he felt a lifetime of sympathy towards the working class. He was drawn to Socialist ideals as a young boy. This NTWU leader took direction from the executive board of the party and relayed those plans to those involved in the Loray Strike. While Weisbord was given respect within his party for his understanding and implementation of Socialist values, the Southern community was not impressed when he came down South and shared these values with them during the Loray Mill Strike. The *Gastonia Gazette* described Weisbord as "an East Side Russian Jew, [who] knows as much about American ideals as Hottentot."[12]

It was not Weisbord, but Fred Beal, a veteran organizer of textile workers from Massachusetts, who initially traveled down South to be the visible physical leader of this proposed Southern mission. Beal was a tall thirty-three-year-old New England man with bright red hair. He had a broad build and youthful features. Beal, like Weisbord, became connected to Communism in his youth. Beal heard the convincing revolutionary leader of the International Workers of the World (IWW or

Wobblies), Bill Haywood, speak in 1912 during an early Lawrence, Massachusetts, strike. Beal himself had been a textile worker since the age of fourteen. After Haywood's speech he soon joined the IWW. Beal was an active and enthusiastic member and eventually became a leader. He led a Lawrence strike in 1923. He later joined other like-minded individuals in Boston in 1927 to protest the executions of Saco and Vanzetti. Beal then led a 1928 New Bedford textile strike. Beal became known for his love of organizing; he joined the NTWU during its conception in 1928, and soon after received his mission to lead the Southern organizing effort.[13]

NTWU Leader Fred Beal (August 1929 *Labor Defender*, restored courtesy Millican Pictorial History Museum).

Weisbord gave Beal the task of establishing the best mill town in which to make a Communist-led stand. Once established in a community, the plan was to fan the union outward to neighboring mill towns. If everything went as planned, in time the whole South would be organized by the NTWU. Eventually Beal became specifically attracted to Gastonia because he believed a takeover here would attract worldwide attention. The community's and mill's status would focus attention and draw funds to the cause. The cause and campaign could then feasibly spread across the country.[14]

As for the Communist roots of the organization, few involved in the NTWU fully understood the complete doctrine and motives of the Communist Party. Few living in the United States involved in this strike had been to a Communist country and actually seen the theories put to test. The word Communism was used by people freely, but few really knew what it looked like. Intellectuals of the day accepted Socialist ideology and colleges exposed students to its theories. Communism attracted a variety of individuals, not only those with an educated background and idealistic beliefs, but workers, minorities, and others injured by the capitalistic system. A classless society with common ownership sounded pretty good to those in the lowest class, working long hard hours and having very little to show from it. In Ella's case and others, hard work didn't even provide for their children.[15]

From the very beginning, funds were tight. Beal rode a motorcycle from New York to North Carolina because it was the cheapest means of transportation. On New Year's Day of 1929, Beal arrived in Charlotte, North Carolina. The only party member in the city, a sympathetic merchant, enthusiastically greeted him. The merchant believed Beal was the textile workers' salvation. Beal initially tried to get a job at local mills, but failed. When Beal asked an owner of a mill for a job, this was the reply:

> No, suh, young man, I'd never take on a Yankee or any other ferriner in my mill, 'n' that thar goes I reckon fo' all the South. They'd put too many strange ideas in the heads of mill-hands—some nonsense lak workin' only eight hours a day. Why, I work nine hours every day 'n' I own this here mill. I guess my hands whould be willin' to work at least 'leven, 'n' by the Lord Jesus Christ, they will! Work never hurt no one. Read yer Bible! It condemned man to hard work forever because he sinned. But nowadays the ferriners, like those Rooshin Communists, are tearin' down religion—but they'll never make headway in the South, because we are all God-fearin' people.[16]

Unable to obtain employment, Beal then focused his time on trying to recruit new members and to get a feel for the South. Many workers were pessimistic about the idea of foreigners coming to the South in order to save them from the mill bosses. Unions had come and gone and the working class had nothing to show for it. When Beal asked about the prospects of unionization, one woman said: "The mill people won't stick together ... no, it's no use, we've tried it before. I've spun, and spun and spun. I guess I've spun in every village in the South, and now, what have I got to show for it?—nothing. The people in the city looks down on us mill-folks. They can always tell us mill-folks from other folks because we's chalky-faced and skinny."[17]

The last statement from the mill hand was reminiscent of previous testimony Beal had heard from the Charlotte merchant union member's son, who said, "You'll have no trouble picking out a textile worker from a human being." Beal later said himself that many of the textile workers looked like "living skeletons." The majority of the middle class in the

"New South" felt more disdain than sympathy towards textile workers. They could not relate to mill folk. Textile workers were demeaned with such derogatory terms as "lintheads" and "cotton mill trash."[18]

Fred Beal felt that it was his mission to help these unfortunates. Beal sought useful information about how to organizing all the textile workers. Worker O.D. Martins told Beal what he wanted to hear when he said that if he succeeded in organizing Gastonia, "you'll organize the South." George Pershing, a fellow NTWU leader, believed that too, later saying "North Carolina is the key to the South. Gaston County is the key to North Carolina, and the Loray Mill is the key to Gaston County."[19]

Before focusing his attention on Gastonia, Beal managed to organize workers at the Chadwick-Hoskins Mill in Pineville. He was able to do this because of the hostility the white workers felt towards the owner's newly acquired practice of hiring black workers. The traditionally white only workforce was outraged that whites were laid off and replaced with blacks. Readily, and in great numbers, the white workers joined the NTWU. The experimental hiring practice did not last long, and black workers were fired and whites rehired.[20]

The factors that enabled Beal to organize his first mill in the South were ironic since the NTWU believed in equality for all, including black workers. In fact the NTWU founders later made blacks' rights a focal issue in the organization's campaign for the South. Among other white workers' complaints at Pineville was being required to live in "nigger" homes. Beal later stated in his autobiography that the Southern workers as a whole were not so concerned with working conditions, or long working hours, as much as the stretch-outs, the meager wages, and the introduction of black workers.[21]

The mill workers' lack of concern over hours and concern over stretch-outs can be explained by the cultural differences between Northerners and Southerners. The Southern culture itself is traditionally a relaxed, laid-back one. Southerners were accustomed to working long hours, but at a slower pace. This way of working goes back over generations of families working all day in the hot sun on farms. The rushed pace to cover an increased work load in a hot stuffy mill was clearly one of the issues that outraged strikers the most. A Loray Mill striker told Vera Buch: "We used to git five, ten minutes off from work now and then. That way you cold bear the mill. Now they keeps you a-runnin' every bit of them eleven hours. Hit just ain't possible for a human bein' to do what they're askin."[22] Beal did not fully understand the Southerners' complaints about the recent stretch-out. He said: "These Southern workers were more interested in protecting their prejudices than their interests. They did not complain about such low wages and long hours as would never be tolerated in the mills of New England but they hated what they called the 'stretch out,' which was accepted as inevitable by the Northern workers." The Southern workers were upset that improved technology did not improve their working conditions, but rather increased expectations put upon them.[23]

The Loray Mill, located in Gastonia, was where Beal went by mid–March of 1929. In Gastonia, Beal saw far worse conditions than he had witnessed elsewhere—conditions he believed seemed primed for a strike. Will Truett was one of the first locals to join the NTWU and organize others. Truett infiltrated the Loray Mill as a worker in order to supply information to Beal. He also helped acquire a small shack for the NTWU which was about a seven-minute walk from the mill. It was converted into the union headquarters. A secret union membership began and a few members joined immediately.[24]

Beal took a trip to New York to establish support and then returned to Charlotte. While in Charlotte he received a wire from Truett. The wire read, "Fired today. What will I do about it?" Beal returned to Gastonia and discovered that Will had been fired for organizing the workers. Beal said, "This is a declaration of war from the bosses, a challenge to us. Unless we take action at once, our union will be smashed to smithereens before it's hardly started."[25]

That night there was a secret meeting held in a textile worker's home. The meeting was described as more of a religious revival by outside Northern NTWU leaders than a structured orderly union meeting. Leaders said that they could barely keep up with those pledging loyalty to the union and signing their names up to join. There was a vote held and an agreement to hold an open meeting the next day behind the mill. This was a daring venture, for up to this point all meetings had been held in secrecy. Mill management had threatened to fire anyone associated with any union. Despite the risk, the group proceeded and was determined to strike if any workers were fired for attending the meeting.[26]

The next day, Saturday, March 30, Ellen Dawson, vice president of the NTWU, a native of Scotland and a young veteran organizer of New England, joined Beal in his efforts. Dawson had experience in strikes in Passaic, New Jersey, where she had worked and in New Bedford, Massachusetts. Dawson immediately took a leadership role next to Beal.[27]

There were several thousand textile workers outside that day as Beal with his Massachusetts accent and Dawson in her Scottish accent urged strikers to stand firm. A few preachers tried to gain the attention of the crowd and persuade the mill workers not to listen to the foreigners, pointing out that they were Communists. Native women textile workers spoke too and testified to the poor mill conditions, stretch-outs, and night work, which reportedly silenced the preachers. Mill supervisors stood nearby recording the names of those workers present. Those in the crowd, especially the women, heckled the bosses with such com-

On March 31, 1929, textile workers in Gastonia vote unanimously to strike (courtesy Gastonia Police Department).

ments as, "What about the stretch-out?" and "How about God and the bathtubs?"—in reference to a remark that a mill-sponsored preacher had made regarding the lack of necessity of bathtubs in mill-village homes because God did not look favorably upon frequent bathing. The taunting eventually drove the mill management to leave.[28]

A handful of workers were dismissed the following day for attending the weekend meeting, and rumor spread throughout the mill that all union members would eventually be fired. The fired day-shift workers met with fellow night-shift workers and union leaders at the newly acquired union headquarters. The decision to meet immediately, openly, and en masse was made.[29]

Later that afternoon, at 3:00 p.m., near some train tracks just outside the mill, Beal addressed the gathered crowd consisting of off-duty night-shift workers and fired day-shift workers. In a harsh, emotional voice, Beal told them in his soapbox address that they could force the boss to give them a fair pay and a forty-hour work week, if they could just stick together. He promised them food and supplies during a strike. Beal and other leaders spoke for about an hour and then called for a strike vote. A sea of mill workers' hands defiantly rose. It was unanimous, and the historic Loray Mill Strike of 1929 began. The impoverished Southern men and women hooted and hollered in the open field in such a manner that the Northern organizers were taken aback. A photograph taken of the auspicious event froze the moment in time. The picture and accompanying headlines spread to international newspapers and magazines.[30]

Earlier that afternoon a handful of men and women were taught the basics of picketing. Directly following the mass meeting the trained men and women and about 150 others led by Beal hiked in pairs to the perimeter of the mill, which was surrounded by a wire fence and gate. Some night workers proceeded to yell to the nearby day shift workers inside the plant. "To hell with the bosses!" "Come on the line!" they yelled. Some workers inside replied back that they wanted to join but that the bosses had locked them in. The strikers, some brandishing signs with sayings such as "Don't be a scab!," "Who has money in the bank? Only the bosses!," and "No union man will starve!" continued picketing for a period of time. Union leaders prompted the singing of the popular union song "Solidarity."[31]

The NTWU distributed circulars prompting all Loray workers, both day and night, to join existing strikers on the picket line. One note read:

To all workers of the Manville-Jencks Mills:

Today union members were fired who have slaved for years in this mill, making profits for the bosses, for no other reason than belonging to a union. Hundreds have already joined the union and the bosses want to fire them, one by one and bring in workers from other states whom they have fooled into thinking that they pay high wages to take our places. All night workers, all day workers STRIKE.

Form picket lines in front of the mill gates at once. Office 1242 West Franklin Street. Every worker out. Clear the mill tonight. Victory will be ours.[32]

Mill management responded by handing out eleven different versions of leaflets, 2,000 of each, discouraging the workers from letting foreigners lead them and proclaiming that the leaders would later abandon them. A sign was hung outside the mill by management that read, "Be a Man—Don't Be Driven." Finding irony in the statement, an older female striker on the picket line haphazardly wrote the same message on a piece of cardboard to represent the reciprocal sentiment of the union and displayed it for all to see.[33]

4

The 1929 Strike

The evening of April Fool's Day, 1929, the night shift didn't enter the Loray Mill and more than half of the 2,200 employees were on strike. By April 3, the general manager, J.A. Baugh, was given a list of demands from the NTWU by a handful of strikers representing the workers. It included: eliminating piecework, hand, or clock systems; the substitution of a standard wage scale; an eight-hour work day; a five-day work week; a minimum weekly wage of $20; the guarantee of equal pay for equal work regarding women and children; appropriate and sanitary housing, including screens and bathtubs; repairs made to mill toilets; a reduction by 50 percent of rent and light charges in the mill village; and recognition of the union.[1]

Baugh expressed sympathy towards the group's desire to better themselves, yet vocalized disappointment and feelings of betrayal towards the committee's actions against management. Baugh, when personally confronted by the strikers representing the NTWU, said: "I feel you have been misguided. We don't object to you people wanting more money and wanting to better your condition. You naturally want those things and we would welcome the opportunity to give you more money if that were possible. I'm more sorry than you ever will know that you have taken this step. I believe if you had considered this from every angle you would have not struck. This is no way to adjust differences."[2]

Baugh quickly dismissed the NTWU demands as being impossible and responded, "You realize that if we should comply with those it would mean that we would virtually give you the plant? You surely don't expect us to do that." Initially Baugh and other Loray Mill officials were unconcerned by the strikers' actions, for there had been strikes before and they had always been short-lived. The *Gastonia Gazette* printed in its April 2 paper that the mill management was "only slightly perturbed" by the disturbance of the day before. On April 3, Baugh said, "I think the situation will be over in a few days. Our home office at Pawtucket, Rhode Island, is not worried at all. There has been no trouble and we do not anticipate any."[3]

Baugh was wrong about the length and involvement in this strike. Immediately reinforcements were coming in. During the second day of the strike, the outspoken George Pershing arrived. He was a party activist sent by both the party's paper, the *Daily Worker*, and the youth organization, the Young Communist League. Pershing was good-looking and flamboyant, the nephew of Edgar J. Pershing, the then chairman of the Indiana Republican State Committee. Pershing's lineage reportedly dated back to General John J. "Black Jack" Pershing, a famous military commander of World War I. George Pershing joined the Communist Party while in prison at Alcatraz. He met party leader Paul Crouch, who was also serving time. Depending on the source, Pershing was incarcerated for being an idealist who had burned down the largest brothel in Honolulu, saving the community from moral ills, or he was a troublesome soldier who, during a night of debauchery, accidently burned down a brothel.[4]

Pershing wasted no time in sharing his and the party's political motives in the South. He announced himself as a Bolshevist sent by the party to help initiate "a gigantic movement in the entire South" with the aim of removing capitalism. He claimed the conditions were ideal in Gastonia for it to be the birthplace for this campaign of organization that would inevitably unite all Southern workers. Pershing further clarified the party's ultimate purpose when he said, "The principal view of the Communist is a control of the country by the workers." The American Communist Party wanted to abolish the wealthy class, which it considered to have too much power in the United States. Pershing further explained that mills such as the Loray Mill, under the Communist plan, would be operated by committee. A representative of each department would participate and the committee would be responsible for selecting a manager who would answer to the committee.[5]

The bold public statements made by Pershing flustered Beal, who believed, and justly so, that the revelations would hinder his organizing efforts. Beal's main concern was organizing the workers of a variety of textile mills in the South, with the Loray Mill being the central focus in an effort to strengthen workers' influence on their own conditions. Beal saw the advertisement of the union's Communist roots and underlying long-term goals as counterproductive to the current task of organizing.[6]

The NTWU leaders were for the most part united in the organization's philosophies, but not uniform with their ideas on how much of that ideology should be imposed on the strike situation, or how soon. Members of the Communist Party arrived in Gastonia, all from the same parent organization, but from various splinter groups, with different roles, agendas, and ideas of how to progress in the South. All party members made the majority of their decisions autonomously, but party officials in New York such as Weisbord had authority over those in the field. For major decisions, or in the case of disagreement between leaders, Northern officials had the last word. They also controlled manpower in regards to how many party representatives arrived in the South and who they would be. Northern party officials determined flow of funds to the Southern leaders, which in turn controlled much of what Southern leaders could accomplish.[7]

Communist Party officials in the North sent over twenty representatives from a variety of related splinter organizations. The following organizations and representatives held a visible presence and influence in the 1929 Loray Mill Strike: the Young Communist League directed programs for children (George Pershing, Clarence Miller, Edith Miller, and Sophie Melvin); the Workers International Relief supplied foods and goods (Amy Schechter and Caroline Drew); the International Labor Defense provided the legal defense and party reporters for the monthly magazine the *Labor Defender* (Carl Reeve and lawyer Tom J. Jimison); and the *Daily Worker* was the communist newspaper (George Pershing). Fred Beal, Vera Buch, and Ellen Dawson were the prominent organizers from the NTWU and led striking activity. Sophie Melvin, Joseph Harrison, and Clarence Miller also held important roles in representing the NTWU.[8]

The members from the various groups arrived sporadically throughout the strike, often stayed briefly, and left as suddenly as they arrived. The directions given by various officials in New York often hampered the efforts of organizers in the Piedmont, who were faced with an unfamiliar and constantly fluctuating situation. Sometimes these orders even contradicted each other. Some Southern-based leaders found themselves disregarding orders from above in favor of their own judgment. This

Oeft to right: an unidentified U.S. Deputy, Carl Reeve, Ellen Dawson, and Tom
Jimison stop to pose during the Loray Mill Strike (courtesy Millican Pictorial
History Museum).

caused tension and conflict between levels and branches of leadership
throughout the strike. There was not a clear and united front within the
party on how to progress.[9]

The topmost chain of command was not Beal and other NTWU
leaders in North Carolina, nor was it Weisbord and other American
Communist Party officials in New York, but rather the founders in the
Soviet Union. Their zeal for rapid social change by the insistence of
racial equality and almost immediate organization of all workers of the
South contrasted with Beal's vision and resulting actions. Due to the
hostility many white workers felt towards black workers, Beal wanted
to sidestep the issue of race for the time being. He wanted to focus solely
on organizing and strengthening the Loray Mill Strike in hopes that this
success would inspire other mill communities to do the same and join
along. He preferred having a strong central thrust rather than a weak,
broad series of strikes. By his own admission, Beal often ignored orders
until forced by Weisbord and other leaders in Northern headquarters
to follow the party's wishes.[10]

Vera Buch arrived from the North to help Beal with his organizing efforts on April 5. Vera, like various other NTWU members, was active in the organization of the textile workers in Passaic, New Jersey, in 1926. Vera was in a relationship with Northern party leader Albert Weisbord and later married him. One of Vera's initial concerns was the low turnout on the picket line. On the first day or so of the strike, there were one to two thousand out protesting, but that number quickly dwindled to a few hundred. As time progressed, representation on the picket line eventually fell to a few dozen women and children.[11]

Mill management and its supporters did not sit idly by. According to NTWU leadership, the local mill owners and supporters met and raised $250,000 to fight the strike. The funds were used to pay for advertisements, guardsmen, and later judicial costs. If this information is accurate, the funds of the mill supporters contrasted sharply with union resources, which throughout the strike were limited. A constant complaint by organizers in the South was insufficient funds. In fact, Vera and Beal both said that at times during the strike there was barely enough to keep the relief store stocked with the bare minimum of food supplies.[12]

Relatively soon after the strike began, a letter from F.L. Jenckes to G.A. Johnstone, who was the manager at the Loray Mill in 1927, appeared. The two-year-old letter most probably was stolen from the Loray Mill office. NTWU leaders hoped publicizing the letter would give confidence to Gastonia residents that what the NTWU said about the greed of the owners and bosses was true. The letter was dated November 8, 1927, and in it Jenckes wrote:

> Dear Mr. Johnstone:
>
> I have been keeping close tabs on your payroll and production at Loray Division and I am glad to say, it is very gratifying to see your payroll come down and your production go up. I am frank to say I was skeptical about your being able to cut $500,000 a year on the Loray payroll and keep your production up. I want to apologize now for this skepticism. Now I think you can cut $1,000,000 a year and still keep production up.
>
> I am in hopes of getting South but you are making such a good job of it that I am only afraid that I will upset things rather than help.
>
> Yours very truly,
> F.L. Jenckes[13]

Public relations work was done energetically by both sides. Leaflets were distributed in the local community from the NTWU listing demands and advertising the following grievances: that male labor worked 60-hour weeks for an average $11 a week pay; women were forced to work night shifts; child labor worked 11-hour days for less than $5 a week; the housing was overcrowded and unsanitary; and workers were forced to do business in the overpriced company store where they were kept forever in debt.[14]

Will Truett, who was now a local NTWU secretary, sent a letter to the *Gastonia Gazette* that was published within the first few days of the strike. Truett reiterated the reason for organizing: "We workers of the Loray Mill are on strike to better our conditions. The only way we can do it is by joining the union so as to be united and to work together in an organized manner."[15]

Alongside the native NTWU member's letter was an editorial titled "Time for Sober Thought," cautioning the community against the NTWU and the threat it held for their way of life: "There is a vast difference between American organized labor as represented by the various unions sponsored by the American Federation of Labor and the 'wildcat' unions sponsored by the Communist Party which is merely another name for the Russian Soviet…. If they [the strikers] will give a second sober thought and take the trouble to discover what Communism is and what it proposes to do we believe the strike situation in Gastonia and in the country generally would vanish in short order."[16]

Printed material from mill owners and their supporters flooded the community. The propaganda's underlying themes included a call to morality and Americanism: "Our Religion, Our Morale, Our Common Decency, Our Government and the very foundations of Modern Civilization, all that we are now and all that we plan for our children IS IN DANGER. Communism will destroy the efforts of Christians of 2,000 years. Do we want it? Will we have it? No! It must Go from the Southland."[17]

Another leaflet said: "Here are the Names of the Union Leaders: Albert Weisbord, Michael Intrater, Leno Cherenko, Peter Russak, Peter Hegelias and Sonia Haress. Are these American Names?" Only Weisbord's name was an actual NTWU leader. The others were fictitious.[18]

April 3 was the first day anti–NTWU information appeared in the *Gastonia Gazette*. The *Gastonia Gazette* was Gastonia's chief newspaper and the most vocal of all the local newspapers about its anti-union and anti–Communist sentiments. It was the newspaper physically closest to NTWU activity, and its customers were directly affected by the events of the strike. Paid anti–NTWU ads appeared in the *Gastonia Gazette* alongside the paper's own unfavorable editorials and articles. This opinionated printed material was seen most often during the infancy of the strike, yet was noticeably visible throughout the striking period. The paid advertisements were marked "Ad Paid for by Citizens of Gaston County." The first of these read in part:

> The very existence, the happiness and the very way of life even, of every citizen of Gaston County is threatened, and is in the balance, if Beal and his Bolshevik associates succeed in having their way. The question in the minds of many people who belong to the Christian church, who belong to the various patriotic and fraternal organizations is: Shall men and women of the type of Beal an associates, with their Bolshevik ideas, with their calls for violence and bloodshed, be permitted to remain in Gaston County?[19]

The following day, April 4, the *Gastonia Gazette* printed a paid advertisement from "Citizens of Gaston County" titled "Mob Rule vs. Law and Order." The ad pleaded again for patriotic community support against the outsiders and their foreign beliefs. In part the article read:

> MEN AND WOMEN OF GASTON COUNTY, ARE YOU WILLING TO PERMIT THE MEN OF THE TYPE OF BEAL AND HIS ASSOCIATES TO CONTINUE TO PREACH THE DOCTRINES OF BOLSHEVISM ANY WHERE IN AMERICA AND ESPECIALLY HERE IN OUR MIDST? ... Before the troops arrived here yesterday the mob was rampant at near the Loray Mill in all of its hideousness, ready to kill, ready to destroy property.... LET EVERY MAN AND WOMAN IN GASTON COUNTY ASK THE QUESTION: AM I WILLING TO ALLOW THE MOB TO CONTROL GASTON COUNTY, THE MOB WHOSE LEADERS DO NOT BELIEVE IN GOD AND WHO WOULD DESTROY THE GOVERNMENT? THE STRIKE AT THE LORAY IS SOMETHING MORE THAN MERELY A FEW MEN STRIKING FOR BETTER WAGES. IT WAS NOT INAUGURATED FOR THAT PURPOSE. IT WAS STARTED SIMPLY FOR THE PURPOSE OF OVERTHROWING THIS GOVERNMENT AND DESTROYING PROPERTY AND TO KILL, KILL, KILL. THE TIME IS AT HAND FOR EVERY AMERICAN TO DO HIS DUTY.[20]

Beal rebuked the *Gastonia Gazette* in his speeches. He claimed that the newspaper was the paid instrument of the mills and not looking out for the well-being of the workers who filled them. The *Gastonia Gazette* retaliated in editorials, such as on April 4, when it printed:

THE GAZETTE WAS HERE FIGHTING THE PEOPLES BATTLES BEFORE BEAL WAS BORN, AND WILL BE HERE WHEN BEAL HAS LEFT THE GASTONIA WORKERS HIGH AND DRY AND HAS MOVED ON TO MORE FERTILE FIELDS, AFTER MILKING THIS ONE DRY.

The *Gastonia Gazette*'s editor also acknowledged in this particular article that it was probable that Loray workers' pay was too low and that too much work was demanded in some departments. However, the *Gastonia Gazette* emphatically believed that improving workers' conditions should not and would not be done by the actions of the NTWU. The *Gastonia Gazette* called the NTWU "a bunch of Russianized, Red, avowed Bolshevists who would destroy capital, the very thing that makes it possible for any of us to work." The *Gastonia Gazette* editorial even went so far as to say: "Whenever the Communists get their bloody claws on America, anarchy will reign and there will be no mills or factories to work in."[21]

On April 5, "A Citizen of Gaston County" published a full-page ad titled "Red Russianism Lifts Its Gory Hands Right Here in GASTONIA" in the *Gastonia Gazette*. The Communist Party's own words found in the *Daily Worker* were used against the NTWU in this advertisement. The *Daily Worker*, the Communist newspaper used to reach Communist sympathizers, was as guilty as the *Gastonia Gazette* for sensationalizing Gastonia NTWU leaders' intentions and potential impact on the community. The anti–NTWU ad found in the *Gastonia Gazette* reiterated the Communist party's conspiratorial agenda: "It is a party that seeks the overthrow of capital, business, and all of the established social order. World revolution is its ultimate goal. It has no religion, it has no color line, it believes in free love—it advocates all those things which the people of the South and of the United States hold sacred."[22]

The ad publicized to the community the party's own statements that Gastonia was merely "the starting point for the building of a communist society." The newspaper also outed that the NTWU as "the American branch of the Russian Communist Party." Another disturbing revelation to the community was that the Communist Party had future plans for white, black, and people of all races to come together socially. A mixed benefit dance in March was being planned. In the minds of communist sympathizers the revelations were signs of progress. However, in conservative, racially divided Gastonia, in which a social and economic crisis was brewing by means of a communist-led strike, the communists' rhetoric fueled loathing.[23]

One unintended consequence of the anti–NTWU articles and ads placed in the *Gastonia Gazette* was that they exposed disenfranchised textile workers to the NTWU. Despite the newspaper's own opinion of the union, the barrage of publicity increased the number of dissatisfied workers joining the NTWU. Other newspapers, especially in the South, offered their own spins on the strike, therefore extending the scope of exposure. Beal spent considerable time speaking to potential strikers, and organizing new locals who had learned about the union through newspaper coverage.[24]

The striking period wasn't all filled with serious business. At points during the strike a festive atmosphere prevailed among the strikers. Men and women who had worked such long hours now had time and energy for recreation. When strikers were not spending time with their families or attending to union business, they spent time socializing. Picnics and gatherings were held and a feeling of warmth, happiness, and hope permeated. The promises of food and support from the NTWU enabled this mood of jubilation. A true transformation took place. When Bertha Hendrix recalled her textile mill experience, this is what she had to say:

> I had been working for Manville-Jenkes Mill in Loray, near Gastonia, for eight years—ever since I was fourteen. We worked thirteen hours a day, and we were so stretched out that lots of times we didn't stop for anything. Sometimes we took sandwiches to work, and ate them as we worked. Sometimes we didn't even get to eat them. If we didn't keep our work up like they wanted us to, they would curse us and threaten to fire us. Some of us made $12 a week, and some a little more.

Bertha then described how she felt when the NTWU organized: "This was the first time I'd ever thought that things could be better; I thought that I would just keep working all my life for thirteen hours a day, like we were."[25]

Workers were optimistic about what the future might bring. There was much to celebrate, and strikers did. The behavior of the newly freed up strikers was looked down upon by much of the community. These townspeople saw their actions and attitudes as inappropriate and it fueled their existing hostility towards them. They thought the strikers were using the NTWU to enable themselves to be lazy and enjoy recreation with no work. A local woman, Mrs. Walter Grigg, expressed her sentiment that the loss of restraint by the young people was what disturbed her most about the strike. She complained about the noise and the open "hugging and kissing" she witnessed.[26]

The behavior of the strikers, especially the young single women, was viewed as being rebellious. A youthful rebellion, a challenge to proper Southern morals, was seen in the women's openly flirtatious behavior and unconventional dress. Both those who appeared too masculine dressed in overalls, and those too sexy, all dolled up, were challenging social attitudes.[27]

The Gastonia youth were part of a larger phenomenon. They were the resulting aftermath of industrialization across the South, the United States, and other parts of the world. Values were changing as quickly as the technology that facilitated change. Farm and factory machinery, radios, cars, and various other inventions enabled change to take place quickly, to the chagrin of those rooted in tradition. Formidable Southern paternalistic beliefs were becoming obsolete as economic, social, and political change took place. The twenties in general were a time of turbulence, transformation, and excitement. The industrialization of the South was fueling massive shifts in lifestyles and morality. The excitement of the strike magnified the feeling that change was taking place.[28]

While Ella May didn't have the means to buy her own radio or car, she was interested in technology. Once when her lover Charlie left his car near her home, without permission or experience, she hopped in and enthusiastically gave herself and her children a ride. It ended with the car crashing into a clump of trees. Fortunately no one was injured by the family thrill ride.[29]

Loray Mill management felt as if they were being encroached upon by the truant workers. They were disturbed by the proximity of the picketing strikers to the mill and adjacent employment office. Management said the strikers were preventing willing workers from entering the plant. Of course, deterring scabs was precisely the strikers' purpose for picketing. Management called upon reinforcements. On April 3, fifty police officers and deputized men attempted to create a buffer zone between the Loray Mill and the strikers. The men tried stretching ropes, which were promptly cut by strikers. Then a cable was strewn across the streets surrounding the mill, attempting to keep the 400 picketers at bay. The strikers pulled and knocked over the blockade, winning a tug-of-war with the cable. One policeman was knocked over during the commotion.[30]

The rope and cable incident on April 3 led to the North Carolina National Guard's being sent in later that afternoon. North Carolina Governor Max Gardner, himself a mill owner in nearby Shelby, called upon the guard after Gastonia's Mayor W.T. Rankin, Chief of Police Orville Aderholt, and Sheriff Eli Lineberger made the combined request. Introducing the National Guard was not simply a method to keep the peace. It was a time-tested means of keeping unions at bay to the point of not being seen or heard. National Guard troops often squashed organizing momentum. Major Dolley of Gastonia was temporarily placed in charge of the incoming guard and after a few days was replaced by General Metts from Raleigh.[31]

After a number of divisions arrived in Gastonia, the streets became vacant, although strikers did not leave complacently. Picketers taunted the guardsmen, calling them "Boy Scouts" and "Nasty Guardsmen." Minor skirmishes broke out as strikers retreated. National Guard reinforcements continued to arrive from neighboring cities, and by night's end five companies and some 250 men were guarding the Loray Mill and patrolling adjacent streets. The units were from Gastonia, Shelby, Lincolnton, and two others from Charlotte.[32]

A strong, feisty woman, Bertha Tompkins, recalled

National Guard Commander Major Dolley and Chief Aderholt (courtesy Gastonia Police Department).

Top: Troops camp outside the Loray Mill to provide 24-hour protection. *Bottom:* National Guard heavily armed (both photographs courtesy Millican Pictorial History Museum).

her experience that day with three or four guardsmen while breaking through their line in front of the mill. "They was pushin' at me, I done tole them, I kin walk, you don't need to push me. That law pushed me again and I had a stick in my hand and I hit him on the head. Then y'all knows what they did, for of 'em took hold of me and dragged me into that wagon and they done put me in a cell in the jail."[33] The guardsmen escorted Bertha as the crowd "gave the party a tremendous send off." Ten individuals in total were arrested, one woman (Bertha) and nine men. The charges were disorderly conduct and one case of carrying a concealed weapon.[34]

A meeting was held between the strikers and their leaders on April 4. Beal, Buch, Pershing, and Dawson explained about the NTWU and its plans to the crowd of 500 or 600. Elation was heard when it was announced that the Workers International Relief would be there within a few days with food and supplies. It was explained that this was the same organization that had raised nearly a million dollars for the Passaic strike. Leaders urged the strikers to be patient and asked that those with the greatest needs be served first when aid arrived. A collection was planned to be taken in town and at other textile plants in order to meet immediate needs.[35]

Early on in the strike Ella May became involved with the NTWU. Ella was not a regular speaker at meetings in the beginning. However, Ella was a regular attendee, both at Bessemer City and in Gastonia. She would stand to the side, listen to the speakers, and observe the crowd. Ella, with her high cheekbones, broad forehead, and penetrating hazel eyes would take in every aspect of the strike. Her intense face was framed in a short brown bob. She would often chew tobacco while attentively watching others speak. Unlike many of the single young women, she did not appear to primp herself prior to meetings. In fact, despite her youthful age of twenty-eight she looked older, worn, and thin. Her plump belly, which contained her unborn child, stuck out against her petite frame. She often held her youngest baby Charlotte in her arms as she listened to strike organizers and fellow workers. While she did not always speak, she did always seem to have a song ready. After union business was done she would perform for the strikers. Her music was personal and emotional, and the crowds loved it.[36]

As the strike progressed and more and more leaders were impris-

International Labor Defense

Membership Card

Name *Ella May*

Branch *Bessemer City*

City *Bessemer City, N.C.*

Date of joining

Sec'y

oned, Ella stepped up and took a more active role speaking to the crowd. She told stories of her own life and sometimes stretched the truth. However, the universal theme of her struggle was felt by the strikers and they could relate to the tattered, frustrated mother. Ella often told one particular story that was the basis of her most popular song, "Mill Mother's Lament":

> I'm the mother of nine. Four of them died with the whooping cough, all at once. I was working nights and nobody to do for them, only Myrtle. She's eleven and a sight of help. I asked the super to put me on the day-shift so I could tend 'em, but he wouldn't. I don't know why. So I had to quit my job and then there wasn't any money for medicine, so they just died. I never could do anything for my children, not even to keep 'em alive, it seems. That's why I'm for the union, so I can do better for them.[37]

Four of Ella's children died within a period of a few years. At least one of those children died from pellagra. Ella's rendition of all four dying at once from whooping cough is much more dramatic. Ella's message of despair and frustration rings clear in this exaggerated version of her plight. Other strikers could relate. They as well as Ella knew there was much to lose if they failed in the strike.

The 1929 Loray Mill Strike was unlike the short term ones that had been previously staged in the area. This strike was getting serious, and both sides seemed tense. Beal was concerned that the situation was becoming too heated. He encouraged young women strikers to flirt with guardsmen, rather than taunt them. He wanted the guardsmen to be unable to shoot upon the crowd if they were called upon to do so. Beal feared violence and made many attempts to avoid it. The striking men wanted to carry guns on the picket line but Beal refused, reiterating that it wasn't a battle. The local striking men believed they needed guns in order to defend themselves. They told Beal that they would be sitting ducks for anti–NTWU violence otherwise. Ellen Dawson also tried to put a halt to the prospect of carrying guns. She explained how, during the Passaic strike, picketers continued on despite police violence. The response from the male strikers was, "Just let us have our guns." As a result of the enforced NTWU weapons ban on the picket line, many men refused to picket, fearing for their lives. As the strike progressed,

Opposite: Ella May's ILD card (November 1929 *Labor Defender*).

physical threats only increased, and as a result public demonstrations by the men further diminished.[38]

Women and children were the overwhelming majority of those publicly demonstrating. The women were seen as less of a physical threat to the men protecting mill interests and therefore less of a target. However, the women did encounter rough physical treatment at times, both on and off the picket line. Arrests were an almost daily occurrence and it was mostly women who filled the local jails. The women understood the risks to their safety and freedom and jeopardized them willingly. They served their time in jail away from their children and would sing to occupy their time. Ella May was one of many such women. The women would sing familiar mountain melodies and invent lyrics in order to create their own unique ballads, describing the events currently influencing their lives. Ella May reportedly wrote several songs while incarcerated. Being locked behind bars and isolated from their children, while emotionally painful for mill mothers, provided a time of reflection from their busy lives to contemplate what they were striking for.

The International Labor Defense (ILD) sent Communist Party member Carl Reeve to provide bail and legal representation to the scores of NTWU members being arrested. Reeve was the son of famous Communist leader Mother Bloor (Ella Reeve Bloor). Carl Reeve was one of the few involved in the strike with close ties to the Soviet Union. He had spent three years there and attended the Communist Party's Lenin School. He had been back in the United States a year before getting involved in the Loray Mill Strike. For two months Reeve stayed in Gastonia, periodically facing authorities in attempts to free strikers and leaders alike. On a number of occasions Reeve claimed that authorities tried to hinder his attempts to bail those charged. Reeve said that property or bonds for bail were denied and that bail was repeatedly raised when cash was at hand. The Gastonia Police Department and even Chief Aderholt specifically were implicated.[39]

Reeve, in addition to being an ILD representative, was the editor of the party's magazine, the *Labor Defender*. Reeve, like Pershing, was vocal about the party's ambitious motives. On at least one occasion Reeve was cornered by angry community members who were outraged by something he wrote in the *Labor Defender*. Reeve had advocated for black equality. Reeve credited strikers, seeing him in trouble in the streets, as having saved him from the mob that displayed a lynching spirit. Due to the threatening actions against him, Reeve acquired a bodyguard, Louis McLaughlin, who traveled alongside him and who was always well-armed.[40]

On one occasion Reeve and McLaughlin were picked up by the National Guard on Gastonia city streets not far from the Loray Mill. Reeve was carrying a telegram from leaders up North with the intention of reading it at a strike meeting. Major Dolley, commander of the guard, read the message, which ridiculed the National Guard, calling them thugs and strikebreakers. Major Dolley, according to Reeve, became enraged and told guard McLaughlin, "You ought to be ashamed to associate with New York foreigners, nigger lovers, and all that trash." Reeve and McLaughlin were released within a short period of time. Soon after the release Reeve asked his bodyguard what he thought about the major's comments. McLaughlin took a moment, spit out his tobacco, and bluntly stated, "I think it was a lot of chicken shit."[41]

Workers had their own personal opinions of union leaders and so did the leaders about each other. Sometimes they were favorable, and other times they were not. Beal, for example, was seen as a weak leader by many supporting the cause. A Loray Mill worker, describing the first time he met Beal, had this to say about him:

> I had heard so much about him by the time I finally met him that I wasn't quite prepared for the man I saw. I expected a huge, powerful man with a fine suit. I expected someone bigger than life, and what I met was a short man wearing a suit that looked twice too big for him and it was all shiny where it was worn out, and his shoes looked ready to fall off his feet. When he spoke, his voice was kind of high-pitched, and he pronounced his words funny. He kept talking about something called a "byce" and it took me a while to realize that he was just talking about a mill boss. And he didn't impress me with anything that he said or did.[42]

Leader Vera Buch too saw Beal in an unfavorable light, believing he was a coward. Vera thought that his lack of physical presence at key points in the strike weakened it. She believed that he put concern for his own personal safety and political aspirations above the needs of the strikers. It is true that Beal spent a large portion of his time away from Gastonia, the central strike, in efforts to organize other mills to keep the NTWU aspirations of organizing the South alive. Beal also refused to picket and demonstrate alongside the strikers, leaving this critical role to the Northern women, often Vera. Beal believed that as leader his role was more behind the scenes. Vera disagreed and thought his decisions sprouted from fear.[43]

While the degree of influence fear had on Beal's visual presence in the strike could be disputed, the reality of the cause of that fear could not. In Beal's autobiography he recalled being arrested and driven around town twice while being urged to leave the county. He received anonymous phone calls that threatened such things as, "Beal, get out of Gastonia within forty-eight hours or you'll leave in a wooden box." "This is a warning." "Out of Gastonia Beal, or you'll be carried out."[44]

Beal and other NTWU leaders were seen by many in the local community as instruments of communism. Communism was an evil word to many Americans and to most Southerners. It was synonymous with the Devil. Communism was seen as an evil force that threatened God-fearing Americans. It was a disease to be squashed before it infested all of America. The Communist Party did not help to suppress the Southerners' fears. The topmost leaders of the party advertised the fact that they wanted the working class united and empowered in the United States. The NTWU's *Labor Defender* even labeled the conflict at Gastonia as the beginning of "class war."[45]

While the Northern leaders understood and believed in the underlying forces behind the NTWU, most of the actual strikers who relied on the organization knew or cared little about them. The textile workers cared about the fact that the NTWU could immediately help them improve their lives. Men and women joined the union so they could feed their families, give their children proper medical treatment, and put clothes on their backs and shoes on their feet. It was that simple.

Many local men and women living in the community were sympathetic to the textile workers' plight. They recognized the workers' poor state and their need for improvement. However, the means by which the workers sought improvement, a union, and even more a communist-driven Northern union led by foreign-born organizers, blinded much of the community's sympathy towards the poor working men, women, and children in favor of self-preservation of their own capitalistic American way of life. There were few local middle-class sympathizers who acknowledged the workers' need for the NTWU.

The community felt threatened, under attack. A war of ideals, livelihood, and autonomy was being fought. The union could potentially have catastrophic effects on the textile industry. The industry was the bread and butter of the county and was the pride of a people who built up a strong industrial success. Gaston County was the third largest county in the state and it had the largest number of incorporated towns or cities. Due to the perceived repercussions, all those connected to the NTWU, union leaders and strikers alike, were seen as a threat. To the majority of the local community, the reasoning behind the textile workers' joining the union was inconsequential when looking at the broader picture of the economic survival of their community.

There was no safe haven for the union men and women. Even churches sided with the mill owners. Ministers denounced the actions of the strikers, denied the need for unionization, and refuted that improved conditions were even needed. At a ministers' meeting during the strike, comments were made about the situation by some of the twelve preachers present: "The quarrel at Loray is between the Northern owners of the mills and some Northern agitators." "Children would rather go to work than go to school." "A larger percent of married women work as clerks in stores at lower pay and with longer hours than in the mills."[46]

Beyond being unsympathetic, many within the local community felt it was their civic duty to stand up against strikers. Loray Mill officials and local civic organizations such as the American Legion organized the Committee of 100. On the record, mill officials stated that the Committee of 100 was nothing more than a group of mill workers trusted with the responsibility of evaluating the character of workers who may be allowed to return to the mill. However, membership of the Committee of 100 extended far beyond Loray Mill employees, and those who joined unquestionably believed their responsibility lay deeper and necessitated them to act with swift and firm action. Much of the local community, including businesses, churches, and the newspaper, supported the group, and mill owners showed unwavering allegiance and support.

The Committee of 100, or the Black 100 as the strikers called them, believed they were protecting the community from communism. Members initiated a number of violent acts against strikers. The committee was strong in both numbers and autonomous power. No power of authority, National Guard troops, or local police, ever hindered their illegal activities. Union members had much to fear from this band of civic-minded citizens. Members of the organization participated in varying degrees of protest against the NTWU and its members, and their allegiance to Americanism was consistent. Cries of "Long live 100% Americanism" would be frequently heard as members tore down tents, kidnapped individuals, and beat those they cornered. As the strike progressed, so did the frequency and intensity of the attacks. Strikers were the easier marks, less protected than their leaders, and the more frequent victims of crimes. Leaders, needless to say, were more sought after. NTWU leaders feared assault to the point that they would move between various homes and hotels in order to avoid detection. Threats against their lives were taken seriously.[47]

Recently deputized men and the Committee of 100 were two different entities, yet many deputized men were also key players in the Committee of 100. Major A.L. Bulwinkle, president and founder of the local American Legion, was the unofficial leader of the Committee of 100. Bulwinkle had previously held office in the United States Congress from 1920 to 1928 and from then on represented the Loray Mill as its legal counsel. When some National Guardsmen were removed from guarding the mill on April 9, it was Bulwinkle who pushed to have American Legion men deputized in order to have replacement guards. Throughout the month of April, National Guardsmen continued to be removed and replaced with Legionnaire men. Inexperienced non-veterans were also deputized on a provisional basis for emergency purposes. Anti-NTWU supporters even frowned upon the character of some of these deputized men, but justified it by saying no others could be found because the pay was so low.[48]

The NTWU's textile strike spread to Bessemer City, where Ella May worked and lived, on April 15. A group of workers on the night shift walked out in the early evening hours at the American Mill No. 1. The American Mill No. 1 employed an unusually high number of black employees, but not a majority as in Ella May's mill, the American Mill No. 2. There was some violence in the form of fist fights as a result of which some of the National Guard stationed at the Loray Mill were called in. The disturbance resulted in five workers' being jailed. Robert Goldberg, mill manager for both the American Mill No. 1 and the American Mill No. 2, stated the next morning that both plants were running at full capacity. There continued to be picketing and demonstrations in front of the mill and deputies continued to guard the facility. No striking activity at the American Mill No. 2 had yet taken place. A demonstration at the Pikney Mill in South Gastonia was also held on the 15th, which led to some National Guard's being called there as well. The following day this plant was also running at full capacity.[49]

A meeting of Gastonia's civic organizations occurred on April 16, drawing about 250 people. The meeting was organized to address the perceived communist problem. The *Gastonia Gazette* printed their conclusions in an April 17 article: "A campaign should be inaugurated to educate our people, workers, merchants, business, and professional men of all walks with the teachings, doctrines, and purposes of the Communists."[50] Speakers at this meeting included Mayor Rankin, General Metts, the Reverend Johnson, Major Bulwinkle, and others. A general meeting for all of Gastonia's citizens was scheduled for April 26. This was to be the "first gun in this campaign." The planned meeting was titled "Americanism vs. Communism." These community meetings were no doubt recruiting grounds for the Committee of 100.[51]

About 100 men raided and destroyed the union headquarters and newly opened relief store in Gastonia late in the evening on April 18. The Workers International Relief had just stocked the store with goods for the strikers the day before. The wooden structure of the headquarters was literally torn to pieces and a heap of rubble remained where the building once stood. The relief store, which was feet away, fared better, being made out of brick, but like the headquarters its plate glass windows were shattered. The attackers stole food and supplies, destroyed the mimeograph machine and furniture, and poured kerosene on the remaining food, which the group had hauled into the middle of the street.[52]

The half dozen union men who had been guarding the store were forced at gunpoint to retreat. The National Guard, which stood 500 feet away at their own tent headquarters, did not react during the incident, and later claimed that they did not hear or see anything. Outside media later declared the National Guard's claims of ignorance unbelievable due to their proximity to the crime and the inevitable noise created by the swinging of axes, sledgehammers and crowbars against the wood and

Destroyed NTWU headquarters (courtesy Millican Pictorial History).

glass of the buildings. Soon after the incident was over, Major Dolley and his men arrived on the scene and proceeded to arrest the six NTWU men, accusing them of destroying their own facility. The destruction of the union relief store and headquarters was one of many violent acts by the Committee of 100. The April 18 aggression was not a complete blow to the striking effort, for it created widespread coverage of the strike. National interest and funds began flowing towards the striking effort as a result.[53]

On April 19 an anti-parade ordinance was passed, denying public street demonstrations without a permit, and by April 21 all National Guardsmen had been withdrawn. It was deputized men alone who became responsible for enforcing the new law and keeping the peace in Gastonia's streets. Forty new men became special deputy sheriffs, and this move frightened NTWU organizers. Pershing said that there was "more likelihood of trouble with the legionnaires who have been made deputy sheriffs and given quarters at the Loray Mill than there had been with the troops." Beal said that it was "another tactic of the mill owners ... to stir up a lynching spirit against the union leaders."[54]

An official copy of the anti-parade ordinance was given to NTWU leaders on April 20. The leaders promptly applied for a permit for that afternoon's demonstration but were denied. The reason for rejection was in part due to "the inflamed spirit aroused among the strikers by the recent attack and wrecking of their headquarters." A pamphlet titled *To the Citizens of Gastonia and Gaston County* was distributed among the community. The pamphlet was designed to inform the community of the new city ordinance. It was signed by Major Stephen B. Dolley, Commanding Provisional Battalion. It read: "I am requesting those persons not having legitimate business in the Loray Mill vicinity to refrain from congregating in and around the mill property, and I am notifying all persons concerned that no unlawful gatherings which in any way will tend to incite the people to a breach of the peace or a riot will be permitted in and around the property adjacent to the Loray Mill, or in Gastonia, or in Gaston County."[55]

A group of strike leaders met in Charlotte on April 21 in order to discuss their future plans. The belief was that the distance from Gastonia would provide them a certain level of privacy and safety, but they were mistaken. At the meeting Beal asked Reeve to mail union application cards from textile workers. As Reeve walked to a nearby mailbox, he was apprehended and taken to the Charlotte jail. Reeve said at the jail he talked his way out of the situation by bringing up his connections.[56]

On April 22, a mass meeting was held for the striking textile workers. Several hundred men, women, and children broke the anti-parade ordinance and proceeded to march en masse from their destroyed relief store towards Loray in order to picket outside the mill. Approximately fifty police officers and special deputies holding pistols, bayonets, and clubs intercepted them and violence erupted. Some of the experienced officers did try to control the situation without violence, but to no avail. Many newly appointed and inexperienced deputies were impulsive and angry, and they administered clubbing indiscriminately. A reporter for the *Charlotte Observer* covering the strike was caught in the middle of the chaos, manhandled, and beaten in the head with a gun butt. Throngs of strikers were injured and twenty-nine were jailed. The next day strikers, this time with half the numbers, tried again to picket and were broken up yet again.[57]

While hundreds of strikers were brutally being denied the right to publicly demonstrate the red carpet was being laid out for thousands in a public demonstration against the NTWU. An open meeting titled "Americanism vs. Communism" was organized on the courthouse lawn in Gastonia on April 26. A speaking stand was built, surrounding streets were roped off for room, and an amplifying system was put in place. Gastonia officials seemed to be going out of their way to halt NTWU public speech while at the same time assisting those civic-minded citizens who desired to speak up against communism.[58]

Not all government officials' responses were unfavorable towards NTWU members. The governor of North Carolina, Max Gardner, did condemn the acts of violence against the workers, specifically regarding the raid on the union headquarters and relief store. In an April 23 *Gastonia Gazette* article he stated that the behavior "deserves and receives my unqualified condemnation." He added, "Such lawlessness from any source is indefensible and will not be tolerated in North Carolina." He ordered Solicitor John Carpenter to "apprehend and prosecute the parties guilty." However, no one was ever charged in connection to the violent acts directed towards workers during the April 18 incident.[59]

Despite the repeated attempts at picketing, and the concerns of the community, the fact was that the NTWU was floundering in its efforts to keep strikers organized and workers out of the mill. Only three weeks into the strike, the *Gastonia Gazette* was reporting that production in the Loray Mill was almost back to normal. The NTWU tried desperately to keep momentum going. They continued organizing demonstrations in Gastonia. National and local leaders met twice daily at 3:00 in the afternoon and 7:30 in the evening. Textile workers across the Piedmont (Gastonia, Bessemer City, Kings Mountain, Shelby, Dallas, Lincolnton, Lowell, Belmont, Charlotte, Pineville, Rock Hill, High Shoals, Lexington, and others) were given pamphlets in their mill communities inviting them to join the cause. Yet striking numbers never reached NTWU leaders' expectations.[60]

April 26 was the first day of the annual convention of the Young Communist League (YCL) held in New York. It was also the day scheduled for the reopening of the striking Pineville and Bessemer City mills by the owners. Ella May's mill, the American Mill No. 2, had just recently gone out on strike. National leaders Pershing and Siroka left the South the night before this scheduled reopening to attend the YCL convention.

The mills reopened as planned and neither leader returned. Vera expressed her frustration at the leaders' actions, saying, "This is always a most critical point in a strike. Then if ever, the organizer must be at hand trying to hold the union people together, to prevent a mass stampede back into the mill." Vera was assigned to take over the two open posts and the organizational efforts at this point were left to only three people, Beal, Buch, and Dawson. Carl Reeve too left, like Pershing and Siroka, to attend the New York convention.[61]

Vera Buch met a small hostile blacklisted group in Pineville during the last week in April. The plant was running normally after only two weeks of disruption and Vera told the handful of ousted individuals that they would receive support from the union, although no more representation at that location would be present on a daily basis. The situation at Bessemer City was little better: the group of ousted workers was larger, near a dozen, mostly women, and included Ella May as well as her lover Charlie Shope. Vera especially appreciated Ella's presence at this arduous time, for she said she was as valuable as any organizer. From here forward Vera made a daily routine of leaving Gastonia in the morning for the short ride to Bessemer City. Vera tried to keep connections with the workers in the mill and encouraged picketing outside. As in Gastonia, the effort seemed futile and the strikers often ended their picketing prematurely.[62]

Ella May, however, was relentless and never gave up hope. She was one of the most dependable strikers. When Ella was asked about why she was striking, she said she was "sticking out for our rights ... giving our children something to live for, so they don't have to slave all their days for nothing like I had to...." Ella wasn't afraid to speak her mind, especially when it came to improving the plight of her family. Ella's daughter Millie recalled, "My mother was a fighter. She was known as a speaker back then, I can remember going to work a couple times and hearing her talk."[63]

Ella attended union gatherings on a regular basis and was Bessemer City's best-known local NTWU leader. She along with other native leaders from the area participated in morning union meetings with strike leader Fred Beal in Gastonia in order to discuss issues concerning workers. Ella's daughter Millie believed Beal was impressed by Ella, perhaps even smitten. He visited Ella at her home more than once. Ella helped with the general bookkeeping for the Bessemer City branch. She was more adept at reading and writing than most textile workers. She took pride in her accurate and organized accounting for the NTWU. In addition, Ella promoted fundraising for the ILD, an organization created to help defend imprisoned organizers and strikers. She also was a determined picketer and was imprisoned at times because of it.[64]

One of Ella May's most significant contributions to the NTWU was her organizing efforts. Ella was instrumental in getting black workers to join the union. She started getting her neighbors in Stumptown to sign union cards early on in the strike, even before union leaders reached Bessemer City. When leaders did make it to Bessemer City to talk to black workers, the workers shied away, but when Ella spoke they listened. They could relate to this woman who lived alongside them in Stumptown. Ella May once said, "I know the colored don't like us but if they see you're poor and humble like themselves, they'll listen to you." She embraced the idea of black workers standing on an equal footing to that of white workers.[65]

Ella May convinced many Bessemer City textile workers to strike, both black and white. When workers in the American Mills didn't show up to work, management tried to convince them to return. Workers told management that Miss Ella May had talked to them and that they wouldn't be returning. One agent of the mill then confronted Ella, likely during picketing outside of one of the American Mills, and threatened that she would get what she had coming. Not only was this woman suc-

Working for the union and I am also Doing all I can for the I L D- and I will continue untill it is through and if you are a worker We want you too for We are going to Have a Union inspite of What The Boss says a striker Ella May.
Bessemer city

This note was found with Ella May's belongings on the back of a union leaflet (November 1929 *Labor Defender*).

cessfully organizing white workers, but black workers too. This was a first and this likely outraged white management, which was not accustomed to having black workers stand up to them. In their minds they likely believed that black workers should be thankful to even have a job in what was considered a white-only industry.[66]

The issue of equality among the races was one of the most heated issues of the strike. This was the first time a strike in the South brought this topic to the table. It not only upset the middle and upper classes, but disturbed numerous textile union members themselves. The idea of empowering black workers to the status of whites was unsettling to a majority of Southerners. The Communist Party supported equality among the races in all ways. Not only equal pay for equal work, but intermingling among the races for the purposes of both work and play was encouraged. No line was drawn. Intermarriage, for example, was an accepted idea, while it was thought of as inconceivable by the mainstream at the time. A number of strikers ripped up their membership cards in protest over these ideas. They who defended the controversial communist-led union against anti-unionists in regards to their foreign roots, ambitions, and anti-religious views would rather return to the poor conditions to which they had been subjected than enjoy better conditions equally with blacks. For some, this explosive issue justified an attack on Ella May. For this woman accepted the idea of racial equality, encouraged it, and implemented it in a way that no one else could. Ella May in this way was truly a woman before her time.[67]

It is hard to understand 1920s white resentment towards blacks without understanding Southern history. The Southern economy after the Civil War went through extreme and difficult change. Emancipation was abrupt. Suddenly thousands of black men and women were competing in the free market. At the same time, Reconstruction transformed the agricultural-centered economy into an industrial one. Racial tension continued and transformed from pre–Civil War days. Each kept to his or her own race, not trusting the other. Blacks and whites separated into their own neighborhoods, schools, and churches, eyeing suspiciously any outside race that wandered near.

Those living in the South at this time believed that while times were trying, experimental change in industry provided the promise and opportunity of unprecedented growth that would benefit both races.

Each year manufacturing, mining, and logging pulled more men and women, adults and children into wage labor. Villages and towns appeared by the thousands where none had appeared before. Both races moved restlessly through the South searching for better opportunity. Plantations and farms largely employed black workers. Black men also worked in lumber camps, sawmills, and steel mills. Black women had opportunities as domestic help. Still blacks faced considerably fewer choices than their white counterparts. White workers were preferred in industry, especially in higher-paying skilled jobs.[68]

Poor Southern whites were ingrained with the fear that blacks were an inherent threat to their livelihood. Bosses unhappy with white production threatened to replace them with black workers. Fear of blacks extended beyond a financial one, to that of a physical one. Whites were conditioned to believe that the black race bred violence. Just about every Southern newspaper issue contained an account of a black wrongdoing, not just local episodes, but far reaching accounts. Those involving blacks from the North were especially sensationalized. Black men were thought to be inclined toward certain crimes, crimes of passion rather than cunning. A native of the North living in the South revealed, "The longer I am here, the more I dread and fear the nigger."[69]

After much controversy and heated discussion, the local NTWU voted to invite black workers to join the union. Leader Reeve proclaimed, "This was the first large Southern strike in which a union insisted on the right of black workers to full equality with white workers as an essential part of its program." Vera, along with Ella's leadership, tried to create some trust between union leaders and black workers in Bessemer City. This was not an easy task due to both the history of race relations in the area and to how the NTWU had handled the issue earlier in its Southern campaign. The first strikers Beal helped organize from the Chadwich-Hoskins Mill struck in great part due to white workers' being put on an equal footing to black workers by the bosses. The NTWU supported these white strikers, and in turn their segregated demands. Pershing too, during the infancy of the Loray Mill Strike, had given in to white workers' pressure to segregate the two races by rope during meetings in Gastonia. Ella at that time was the sole striker to cross the segregated line.[70]

The top communist leaders involved in the strike emphasized the subject of black equality. Weisbord advertised in a union meeting: "Every man and woman white or black, tan, yellow, or red that comes into this organization comes in an equal footing." Against Beal's own wishes he conceded to his superiors and announced in meetings that "there must be no division between white and colored workers." May 12–19 was National Negro Week and the union increased efforts at this time to promote black recruitment and break down racial prejudice. The NTWU shared plans and progress for the week with the press. The *Daily Worker* printed: "National Negro Week has been set aside by the Communist Party as a period of intensifying the drive for drawing new negro workers into the party.... We are mobilizing the entire party membership to concentrate on negro workers, especially in proceeding with the various organization campaigns...."[71]

While the American Mill No. 2 in Bessemer City had a large population of black workers, the majority of the mills organized under the NTWU did not. The enormous Loray Mill, for example, with over 2,000 employees, had only a handful of black workers tied to menial tasks. The NTWU made attempts to recruit local Southern black members regardless of occupation. Laundry workers, servants, cooks, and mill operatives were all approached. Time and time again Northern organizers like Vera Buch tried to reach out and connect with black workers to no avail. Vera recalled talking to a black barber about joining the union, one of the few blacks who would respond to her. Vera said: "He, too, wouldn't look at me, with his eyes on the ceiling he said something like this: 'No'm, I didn't get to see that man Miz Ella May done tole me about. But we's gon' have a meetin', shoh enuf.'"[72] Vera expressed her frustration when she tried to communicate to a small group of black women in Bessemer City. She recalled: "It was strange talking to people who wouldn't look at me. Not one looked up from her work or gave any sign she knew someone was talking to her about a union that was for all the workers regardless of skin color, that might help her get more money."[73]

Ella May was the one organizer who connected with and could relate to black workers. She saw their plight as the same as her own, and they saw that indeed it was. She worked with them in an integrated mill, American Mill No. 2, and lived with them in a black mill community, Stumptown. Ella kept an eye out for her black neighbors and they did the same for her. A teen-aged black neighbor would visit Ella's children from time to time while Ella was away, sent by her mother to see if there was anything she could do to help. Because Ella and her black neighbors shared experiences and difficult circumstances, they were allies, and even more than that, friends. Ella had no more than they did. She had no airs about her. She said it like it was, to the point, with no ideological fluff. Her credentials to speak did not lie in a high position she held somewhere, but rather due to her current lowly status. Black workers listened to her like no one else. They liked what she had to say, and as a result many of them joined the NTWU.[74]

Once Ella had enough union cards signed by black workers to justify a meeting, she arranged one for the newly joined black NTWU members. She managed a time for Vera Buch and Albert Weisbord to speak in Stumptown. A wooden box was placed on the ground and about fifteen black men meandered around the area as Vera and Weisbord arrived. Weisbord first took the stand as the men cautiously eyed him from about thirty feet away. He urged them to come closer and they did so, but gradually, as he spoke. Vera addressed the crowd after Weisbord and by the time she finished speaking the two were surrounded, encircled by smiling faces and outreached hands. Vera was especially moved by the experience at welcoming hard-won black members to the NTWU.[75]

The party's paper, which was always extremely optimistic, continued to tout the idea of a future racially unified working-class utopia. A *Daily Worker* article stated: "A common destiny awaits both the black and white worker of the South, both the black and white farmer of the South, a workers' and farmers' government, under the leadership of the Communist Party, established through a dictatorship of the entire working class of the South, both black and white."[76]

Racially themed party propaganda was fuel for NTWU opponents. Mill supporters jumped upon the opportunity to attempt to further ostracize the community from the NTWU and draw an apparently insurmountable divide between the organization's leaders and the poor Southern workforce they represented. Deep racial discrimination of the time made the statements about communist racial ideals disturbing to just about any white, Northern or Southern, male or female, working class or elite, educated or not. Anti-NTWU ads and articles placed the racial issue forefront in the minds of the Southern community. The rhetoric included: "Would you belong to a union which opposes white supremacy? Would you want your sister to marry a buck nigger?" These questions from mill supporters were put during a time in history when segregation everywhere in the South was considered the only way of life.[77]

Northern blacks came down to help recruit Southern blacks. These daring volunteers, however, spoke primarily to white audiences. Gathering a black audience for an unfamiliar Northerner, even a black one, was very difficult. John H. Owen, a Northern black NTWU member, made a brief appearance in Gastonia. White organizer Caroline Drew, part of the Workers International Relief, introduced him publicly to a crowd of strikers as her brother. This made him very uneasy, for he knew this was not culturally accepted in the South. He felt uncomfortable during his entire stay, feeling as if he was being watched at all times. Eventually his white escort, a local striker, abandoned him out of fear of violence. He gave warning to Owen that he was "due for a bullet at any time and no investigation would be made." It didn't take long before Owen himself made a hasty retreat out of the South altogether. He was only in the area for a few days before the foreign hostile land was too much.[78]

A more hardened veteran Northern black organizer, Otto Hall, was sent to replace Owen, but he too felt the danger and only stayed briefly. Hall had a reputation for standing up boldly, some would say foolishly, against white repression. Apparently this was an inherited trait. His family had fled the South years earlier after his grandfather killed a Klansman who had forced his way through the family's cabin door. Hall, with his strong belief in equality, was embraced into the Communist Party and became a member of the party's Central Committee and Negro Department.[79]

Hall, like Owen, quickly raised eyebrows due to the open, public affection paid to him by a white female organizer. Northern organizer Sophie Melvin, without thought of her surroundings, embraced him warmly on the busy public Southern streets of Charlotte when he first arrived. Hall stayed on despite uncomfortable tension until retreat seemed like the only option. On the night of the police chief's shooting, the streets were filled with individuals seeking revenge. History had taught the Southern strikers that when vengeance is to be had, black men were often the first targets. Strikers sought out Hall to warn him of the danger. Hall was found on his way from Bessemer City to Gastonia. He, like Ella, had likely been watching a party propaganda film that played in Bessemer City that night. The strikers quickly transferred him to their car and transported him to a Charlotte train station. The NTWU's *Labor Defender* claimed that Hall's retreat from the South, hidden in a rumble seat of a Ford, was the first record of white men saving a black man from lynching.[80]

All attempts to bring in outside black party members to rally workers' sympathy towards racial equality were abandoned. Despite party leaders' wishes, black strikers were never in great numbers or completely put on an equal footing with their white counterparts. However, there were black workers with union cards, and this itself was a great accomplishment. The credit for this goes to Ella May. She could gain their trust and open them to the idea of organizing. This ability to effectively communicate with others was noted by NTWU leaders and called upon again.

Ella and a select few were given the opportunity to share their textile working stories outside the borders of Gaston County in early May. A delegation consisting of nearly a dozen strikers and leaders gathered into two cars to go to Washington in order to seek the attention of the Senate Committee on Manufacturers in order to support Montana Senator Burton K. Wheeler's proposal for a federal investigation into the Southern textile industry. The strikers participating in the excursion brought past paychecks as proof of their below-industry reported pay. The recent outcropping of strikes in North Carolina, South Carolina, and Tennessee prompted the federal probe. The trip to the nation's capital was probably extremely exciting for Ella, for she had never experienced any life outside of the rural Southern mountains, plantations, and mills. Ella's fiery personality was not tamed by the experience, and she was quoted across the country for what she said to one particular legislator during an encounter on the trip.[81]

The group included Ella May, Kelly "Red" Hendricks, and Binnie Green, and was led by outspoken party leader Carl Reeve. Red was a young, tall, pale, thin man with red hair. He often spoke at union gatherings. He had previously been beaten with a blackjack by deputies during picketing at the Loray Mill. Later he was tried with other NTWU members for the murder of Police Chief Aderholt.[82]

The NTWU delegation that traveled to D.C. Front, left to right: unknown, Ella May, Binnie Green, unknown. Back, left to right: Kelley "Red" Hendricks, unknown, Carl Reeve, unknown, unknown, Tom Jimison (courtesy Millican Pictorial History Museum).

Binnie was a tiny emaciated fourteen-year-old girl whom the *New York World* called "Textile Mill Waif." Binnie was said to look no older than ten. Those in Washington who encountered her were in disbelief that she didn't go to school and worked in a mill. Binnie worked at the Loray Mill up until the strike. She was what the mill called a temporary worker, yet had been so for two years. Despite her experience her title allotted her to make only half as much money as a full-time employee. On average she took home a weekly wage of $5. At one point of Binnie's employment she was an apprentice learning to run machinery. While an apprentice, Binnie made merely 25 cents for two weeks of employment. Binnie and her younger brother were the breadwinners of a family of five. Another brother had died of tuberculosis after working in the mill for several years. Binnie was a good-natured girl whose upper lip often uncharacteristically stuck out from chewing tobacco.[83]

Overall, the delegation did not have much success in Washington;

many senators were unavailable, and others just flatly refused to meet with the small ragtag group led by communists. The Senate committee met to discuss the current situation with the textile mills of North Carolina, South Carolina, and Tennessee on May 8, 9, and 10. The committee unexpectedly broke the session on May 10, which incidentally was the date the NTWU group arrived in Washington. As a result of the abrupt adjournment, the NTWU group did not get a chance to speak in front of the Senate committee.[84]

The American Federation of Labor (AFL) did have an opportunity for its textile workers to speak prior to the break. The AFL was responsible for initiating the hearing, having members of the United Textile Workers involved in the large strike in Elizabethton, Tennessee. They had strikers in Southern textile plants share personal testimony as to conditions, hours, child labor, and pay. During testimony before the committee the AFL president blamed mill conditions for "sowing the seeds of communism." The AFL, which detested the NTWU's communist affiliation, repeatedly made it public in 1929 that they wanted nothing to do with the NTWU and their motives. NTWU leaders blamed the AFL for their cold reception at the capital. The AFL president did indeed fuel the fire of resentment towards the NTWU group, but many senators had already formed their own disagreeable opinions of the organization led by vocal, radical communists.[85]

Despite initially avoiding the group, Senate committee members Wheeler (D–Montana) and Robert M. La Follette, Jr. (R–Wisconsin), relented and did eventually meet with the NTWU delegation privately before they left Washington. Senator LaFollette, chairman of the committee, claimed to have no knowledge of the NTWU delegation's plans to speak. Senator Wheeler told the delegation he was unfamiliar with the signature on the telegram he received, and forgot about their intentions altogether when they did not arrive when he expected. The two senators listened to the strikers, expressed sympathy towards their plight, but explained to them that because some senators had already left the area, the meeting could not be continued. The Senate committee investigating Southern mills reconvened on May 26, long after the NTWU strikers had left.[86]

The NTWU group also planned on speaking at the Women's Trade Union Leagues (WTUL) annual convention, but the organization refused to let them speak due to their connection to communism. A member of the WTUL's board of directors, Ethel Smith, said that it was "unfortunate that the NTWU was under communist leadership." A handful of delegates did talk directly to the strikers outside of the convention despite disapproval from fellow members.[87]

One moment of victory occurred while the group meandered around the capital. The strikers ran into the junior senator from North Carolina, Lee Slater Overman. Overman vehemently opposed the investigation into Southern textile mills, citing throughout the hearing process that it would just encourage the communist agitators, whom he believed were the real problem. He was escorting a delegation of young women from Greensboro's University of North Carolina Women's College when he was intercepted. He was unable to avoid the NTWU group and the experience was no doubt uncomfortable for the senator. The well-educated, well-dressed young women of his state, whom no doubt he wanted to impress, contrasted sharply with the shabby crew of North Carolina textile strikers. The senator told Binnie to get back to school and an awkward silence followed. Ella May then confronted the senator. Ella, with hands on hips, demanded, "How could mill children go to school? How can I send my children to school when I can't make enough to clothe them decently? When I go to the mill at night I have to lock them up at night by their lone selves. I can't have anyone to look after them. Last winter when two of them were sick with the flu I had to leave them at home in bed when I went to work. I can't get them enough good clothes to send them to Sunday School."[88] The senator excused himself as soon as possible with the "glorious girls" in tow.

The press seized upon the media opportunity that the accidental encounter allowed. The story made national news, and soon journalists were seeking the group for similar stories of deprivation. Carl Reeve said the incident was "a breakthrough in the national coverage of the strike." The encounter did not go without criticism. North Carolina Senator Furnifold Simmons said the group had been "very carefully dressed in their poorest garments." Later Senator Overman invited the press to see a group of North Carolina textile workers that he proclaimed had not been strategically chosen by communists. It is true the NTWU group looked destitute, so much so that a Washington reporter pulled Reeve aside to hand him $20 to feed them. Additional reporters followed suit and made their own donations. So while the NTWU delegation was not recognized by most in office, the press certainly acknowledged their plight. They embraced the strikers on both personal and journalistic levels, and textile workers' suffering was brought to light.[89]

Another breakthrough moment of the trip occurred at an assembly of the National Negro Congress. At this meeting whites and blacks sat side by side discussing labor and racial issues. This was a striking change to the segregated meetings that were held virtually everywhere in the South. No doubt Ella, as she did in Stumptown, mingled with ease between the races. The experience of blacks and whites working so closely together for a common cause encouraged the NTWU delegation and raised their spirits. On the ride home the delegation chatted excitedly about what they had just witnessed.[90]

The two most significant achievements the NTWU delegation managed to pull off in D.C. were the participation in an integrated meeting with the National Negro Congress and the launch of national news coverage for the NTWU. Ella May was central to both those developments and therefore key to the NTWU's success in Washington, D.C.

Ella May did not shy away from attention or controversy. She was clearly independent-minded, bold, and vocal. She came from a line of strong women found in the Appalachian Mountains. She and other women textile workers who had originated from independent farms in the highlands were accustomed to being self-reliant. They exuded an air of pride, resilience, and resoluteness. Not all female textile workers demonstrated these qualities. A generation away from the mountains often made women more tame. As NTWU leader Vera Buch observed, after decades of work in the mills, the fiery spirit was beaten down with time.[91]

It is important to note that it was atypical for women to be engaged in industrial work at this period of time. Women worked, but their daily toil was centered on the home with child rearing and family maintenance. Women on farms, such as Ella in her youth, were accustomed to additional labors such as tending animals, planting, and harvesting. This work, while expected to be done, was not credited as being real work. Real work was considered to be those labors that supported a family monetarily and that were performed outside of the home. Prior to World War II, the only women who took on tasks outside the home were those who were extremely poor or those whose husbands were the victims of illness or injury. Ella had the misfortune of both.[92]

While homemakers have been generally forgotten by history, working women's contributions have been underreported, including those working women who were agents of change in the workplace. Both native women such as Ella May and NTWU female leaders such as Vera Buch took on the brunt of the work during the Loray Mill Strike. Women were the most numerous as well as most publicly active participants in the strike. The *Gastonia Gazette* called the native women "the most outspoken and determined" protesters. Following an outspoken meeting, the *Charlotte Observer* printed, "If Gastonia has never realized that militant women were within its bounds it certainly does now." Ella May was the most recognized native woman active in organization efforts of the strike, but she was only one of many women who played key roles.[93]

Striker Daisy McDonald was one such individual and her story was quite similar to that of Ella May's. Daisy was a tall, dark, slim woman who emanated an air of strength and dignity. Like Ella, she was part Cherokee, originally from the mountains, and the sole breadwinner of her family consisting of seven children. Her husband was unable to work for he suffered from tuberculosis and had only one leg. Daisy made $12.90 a week untangling yarn. She had spent 20 years in the mills doing various jobs such as spooling, warping, carding, and sorting. Daisy said she never learned to do the most profitable job, weaving, because while learning the skill she would not have been paid and her family would not have survived without the income. Daisy was one of a handful of balladeers who would sing music from time to time at union meetings throughout the strike. Two of Daisy's songs that survived include "The Speakers Didn't Mind" and "On a Summer Eve."[94]

Daisy was a fierce woman who wasn't afraid of confrontation. During eviction from her mill home, she threatened to shoot anyone who dared handle her elderly mother roughly or rush her family outside. Her unapologetic loyalty and dedication to the NTWU resulted in much suffering for both her and her family, including harassment and the arrest of her husband. In appreciation for her dedication and sacrifice, the NTWU arranged for Daisy's eleven-year-old son Elmer to attend a Young Pioneers camp located in the Soviet Union.[95]

Elmer was one of six young men sent to the Soviet Union from the

United States to participate in the Young Pioneers Convention of 1929. The convention included Young Pioneers from Germany, England, the United States, Canada, and China. Most of the 42,000 children present at the convention were natives of the Soviet Union. All Young Pioneers were well versed on Soviet progress, plans, and values. Elmer found that his hometown of Gastonia was well known by both adults and children there. Shelley Strickland, a twelve-year-old black Philadelphian boy, was struck by his warm welcome. He was surprised that the white people he met treated him as well as any white child.[96]

Elmer's mother Daisy worked closely with nineteen-year-old Northern organizer Sophie Melvin during the strike. Melvin was described as pretty, strong, very capable, and eager to work. Melvin had hitchhiked from New York in response to a call for party volunteers. She herself had experience working in a factory and was active in the Young Communist League (YCL) from the age of fourteen. At the age of sixteen she helped organize a children's playground during the Passaic Strike. Her job in Gastonia was also to ease the burden of mothers by providing child care.[97]

Sophie Melvin and twenty-year-old Edith Miller (also from the YCL) organized a parade for the textile strikers' children on May 30. Although the march was held away from the Loray Mill, was not technically a picket line, and was held by children, it was broken up. About half a dozen children were arrested and held overnight in jail. While Melvin typically worked with children, Miller usually worked with adolescents, holding classes that discussed political and economic issues. Miller was a sophisticated, educated organizer who, like Melvin, had experience in the Passaic Strike.[98]

Another noteworthy woman involved in the strike was Amy Schechter. She was an unkempt thirty-seven-year-old Northern organizer from a privileged background. Her roots were from Cambridge, England, and her father was an Oxford professor. She joined the British Communist Party in 1920 and the American Communist Party in 1921. She was described as having a cockney accent. Schechter was against traditional social convention and embraced rebellion. She had a repertoire of songs she loved to share. Phrases such as "Oh girls, oh girls, take warning and never let it be. Never let a sailor go higher than your knee," would be sung to the amusement and delight of fellow women. She was sent by the NTWU's Workers International Relief to direct relief efforts in the Loray Mill Strike.[99]

Initially vice-president of the NTWU, Ellen Dawson was at the forefront of organization efforts. Ellen was a petite woman with a strong personality. A native of Scotland, she was a rousing speaker and a veteran organizer of a number of Northern strikes. She had worked as a textile worker in New Bedford, Massachusetts, before becoming a striker and later an organizer. She arrived on March 30 and spoke to the mass of workers gathered that day with Beal, and remained a pivotal speaker until her abrupt departure on April 17. She was arrested on a New Jersey

Some leaders and strikers who played key roles during the strike. Back, left to right: unknown, Daisy McDonald, Paul Crouch, Tom Jimison, Carl Reeve. Middle, left to right: unknown, unknown, Kelley "Red" Hendricks. Front, left-right: Kermit Harden, unknown, unknown, Ruby McMahon, unknown (July 1929 *Labor Defender*, restored courtesy Millican Pictorial History Museum).

warrant for apparently violating federal immigration laws. The NTWU declared these charges trumped-up. She was bailed out of jail, but had to remain up North to defend herself. She left the strike right before violence erupted and never returned.[100]

Arguably the most influential woman involved in the Loray Mill Strike was NTWU organizer Vera Buch. Vera and Ella became close friends as the strike progressed, each holding mutual admiration for each other's gritty determination. Vera's connection to communism began as a young adult when she spent time in a New York sanatorium. Vera, suffering from tuberculosis, met another young woman at the sanatorium who was a Socialist Party Member. Vera, having a background of poverty herself, began to develop an analysis of the roots of poverty. After she recovered from her illness she joined the Socialist Party and the Industrial Workers of the World. She joined the Communist Party in 1919 when it formed, and later the NTWU during its conception.[101]

Vera had been involved in strikes prior to Gastonia, but had only played minor roles, such as an office secretary during the Passaic Strike of 1926. In the Passaic Strike, while working for the union, she met and worked with the leader of the strike, Albert Weisbord. Weisbord and Buch remained companions and were married following the Loray Mill Strike. Weisbord was the national secretary of the NTWU by time the 1929 strike materialized. Vera was thrust into being a major organizer at this time, and as a result her responsibilities and tasks increased. Once Dawson left, she and Fred Beal were the major leaders of the Loray Mill Strike. As the strike became more heated, Beal's leadership fell to behind the scenes and Vera remained in the thick of things in the streets. Her understanding and frustration over the gravity and futility of the Loray Mill Strike situation grew as the strike progressed.[102]

A blow was struck to the NTWU on May 6 when it was announced by the magistrate that strikers in Gastonia would be evicted from their company homes beginning the next day. The *Gastonia Gazette*'s May 6 article proclaimed that the former Loray Mill employees "must vacate the houses, unless bond is arranged to cover rental for a period of a year." Initially 62 strikers, 30 families, were ordered to leave. For the most part the evictions went peacefully and orderly. An early exception to this occurred on the second day of evictions when deputies met a family by the name of Robinson who barricaded themselves in their home, bracing doors and windows. Despite threats made by the family that they would shoot the deputies if they forced them from their home, deputies managed to push their way through without violence. Eventually, about a thousand people from two hundred homes were displaced and their belongings littered the streets. According to the *Gastonia Gazette*:

> The Loray Village presented a rather disorganized appearance ... with quantities of furniture adorning spots in most every block in the section. Chairs, beds, tables, and all household goods were deposited on the sidewalk in a haphazard manner.
>
> With no place to go evicted strikers congregated in the streets, sitting on their furniture. One family even used their range in the middle of the street.[103]

By May 16 the union had acquired some land outside Gastonia and built a tent colony for evicted strikers. A headquarters and relief store were hastily built as well. Locals, possibly members of the Committee of 100, taunted strikers as they rebuilt. They threatened to destroy whatever was constructed within three days, burning it to the ground. Union members met and voted to organize a guarded defense in order to protect their property. The same day their infrastructure was restored, their

Strikers' tent city in Gastonia (courtesy Millican Pictorial History Museum).

NTWU members pose in front of their newly rebuilt headquarters and relief store (July 1929 *Labor Defender*, restored courtesy Millican Pictorial History Museum).

leaders sent a letter to the governor stating that since the union cannot rely on others for protection, they would protect their own lives and property "at all costs." The letter read:

May 16, 1929

Max Gardner
Governor of the State of North Carolina
Raleigh, NC
Sir,

The textile strikers of Gastonia are building with their own hands new union headquarters to take the place of the one demolished by thugs while the state militiamen were looking on. The new building is about to be finished and the dedication will take place next Saturday evening, May 18, before thousands of workers.

It is rumored in Gastonia that enemies of the workers, inspired by the mill owners, are plotting to wreck our new headquarters within three days after completion. The Strike Committee took the matter up today and decided that it is useless to expect the one-sided Manville-Jenckes law to protect the life and property of the many striking workers of Gastonia. Every striker is determined to defend the new union headquarters at all costs.

Very truly yours.
Roy Stroud,
Chairman of the Strike Committee

The new relief store and headquarters was built alongside the tent city (courtesy Gastonia Police Department).

According to Beal, police deputy Tom Gilbert was present May 18, the day of the dedication of the tent colony, and made a side comment to striker Dewey Martin: "You'd better make the most of it now, big boy, because it won't stand a week." The deputy's comments, like countless other threatening ones thrown to NTWU members, were not seen as ethically wrong by mill supporters. Gilbert and others justified violence when they saw those they opposed as a communist threat.[104]

The sense of looming violence was tangible. Every striking textile worker and every member of that textile worker's family felt it. Ella May's children remember their mother being so concerned about possible assaults that she felt uncomfortable having her back to the windows and door of their home. The cause for alarm was warranted. Ella's eldest daughter Myrtle was just eleven when she was raped during the 1929 reign of terror. She was attacked by an unknown man in her own home while her mother was away. She was forced back into a room and barricaded in, and her younger siblings futilely tried to come to their big sister's aide as she called out for help. The younger children were unable to open the door that blocked them and stood there helpless, looking through knots in the door as their sister was being raped.[105]

In the eyes of many, Ella May was a loud-mouthed linthead whose unrestricted voice was bringing national shame upon the community. Even more than that, many believed her actions were rallying up scores of others who threatened their God-fearing way of life. Worse yet, folks were disgusted by this white woman who mingled with and tried to cause an uprising among black workers. Ella May was hated.

Another assault on the family was made when their water supply, a nearby spring running into a small stream, was poisoned. One evening, a man arrived at Ella's doorstep at midnight claiming to be interested in joining the union. He left without incident, but the next afternoon Ella noticed when she went to fetch water that it had changed to a strange blue color and had a foul smell. The NTWU had the spring tested and contamination was verified. The poisoning occurred the night prior to a union meeting in Stumptown.[106]

A few days later the same man was apprehended by strikers for trespassing at the NTWU tent camp. He was walking near a stream that was the community's water supply. The man was carrying vials of liquid and a blackjack. Strikers held the man at union headquarters until police officers arrived. Police Chief Aderholt eventually arrested the poisoner. Loray Mill management paid bail for the man and he was back on the streets within hours. Mill management and law enforcement seemed to be working hand-in-hand with and enabling those who wanted to physically harm the men, women, and children associated with the NTWU. They were allowing violent criminals to roam free. Repeatedly strikers were physically assaulted with no one held responsible. Not a single person during the Loray Mill Strike was ever found guilty of a crime against a striker or organizer. Ella and the others holding union cards knew very well what was happening, that their lives were constantly at risk.[107]

A renewed effort was made to persuade workers who had slowly returned to the Loray Mill to walk out again. A secret meeting was held on June 1 in which some workers who were currently working in the Loray Mill plotted with Beal to arrange another mass walkout. On Friday, June 7, around 8:00 p.m., shortly before the 9:00 p.m. walkout, union leaders addressed the mass of strikers gathered at the tent colony. Vera explained to the assembled group their picketing plan to support those

4. The 1929 Strike

leaving the plant that evening. Supportive strikers were not the only people in the crowd. Angry anti-unionists were present as well. As Vera spoke, an egg suddenly smashed into the wall behind her, then a volley of eggs and rocks followed. Beal took the stand and threatened the antagonists that if they kept it up they would get as good as they gave. While addressing the troublemakers, Beal spotted a man pointing a pistol directly at him. A moment later a striker grabbed the gunman and a shot was fired into the ground. Anti-unionists laughed. Police officers at the scene laughed as well. Jeering from the anti-unionists brought the meeting to a halt. Officers did not intervene. Eventually many hecklers were forcibly removed by NTWU strikers themselves. Strikers pleaded with the police to arrest those who caused the most violent disturbance. Eventually, upon insistence, officers removed anti-unionists. The gunman who fired, Hanna, was one of those identified and turned over to the officers. He was never charged with any crime.[108]

Following the assembly a group of mainly women and children paraded down the streets towards the Loray Mill in an effort to support those workers planning on leaving the plant. That particular day and time was picked because it was pay night and men and women would have received their pay envelopes. Beal heard a rumor that the mill bosses had caught on to the plot, yet plans proceeded as scheduled. Company men, special deputies, and local police officers were waiting, and violence erupted. According to Vera, the following happened to her: "One of the cops, a huge burly man, advanced toward me, cursing. His eyes were bulging, his face was red, and he glared at me hatefully as he uttered those obscene words. Then he raised his arm and with his big hand grabbed me by the throat, squeezed it, and shook me."[109] Others caught up in the mayhem received similar treatment. The strikers never made it to the mill. They retreated back to their tents and headquarters.

The violence was not over. That night tragedy struck Gastonia. By the end of the evening the police chief would be mortally wounded, scores of NTWU members would land in jail, and a mob of hundreds would be roaming the streets seeking violent retribution. The police chief was Chief Orville Aderholt. He was the mild-mannered police chief of Gastonia and was looked highly upon by his community, including many textile strikers. Some of his officers, on the other hand, were not held in such high esteem. One of these officers with questionable

Police Chief Orville Aderholt (courtesy Millican Pictorial History Museum).

character contributed to the police chief's demise. Officer Tom Gilbert had a history of being hostile towards the strikers and he was up to debauchery on June 7.[110]

Gastonia deputy Gilbert and his friend Arthur Roach, a former officer, took a trip that day to nearby Charlotte, where the Grand Parade of Confederate Veterans was being held. The two deputies watched the parade and drank. On their way back to Gastonia they stopped at Pedro Melton's filling station looking for more alcohol. Upon being told they would not be given any, they chased the owner of the establishment at gunpoint into the nearby Catawba River. The officers

Deputy Tom Gilbert and former officer Arthur Roach (courtesy Millican Pictorial History Museum).

114

fired several shots at the man and then turned on his friend J.C. Hensley and beat him. Two county police officers reached the scene and told them to "return to the Gaston side of the river." The angry drunken men took their frustration and wrath and redirected it towards the NTWU. They were already enraged with the knowledge that union activity had taken place earlier that day in town and so decided to enter the union camp. It is uncertain what exactly took place at that point due to contradicting accounts. However, union guards were armed at the camp and keeping an eye out for attackers when the two approached.[111]

Deputy Charles Ferguson (courtesy Gastonia Police Department).

At 8:45 p.m. the phone rang at the Gastonia police station and Officer Charles Ferguson picked up the phone. A woman reported a disturbance at the tent colony and Chief Aderholt and Officer Ferguson left to investigate. Most likely the chief and accompanying officer went to the camp to keep peace between the drunken deputies and the on-edge union guards. In the unclear moments that followed, shots were fired and Chief Aderholt was critically injured, shot in the back and lungs. The other officers on the scene, Roach, Gilbert, and Ferguson, were also injured by gunfire. Joe Harrison, a member of the NTWU, was shot. However, none shot during the melee were as seriously injured as Aderholt.[112]

The next morning Chief Aderholt died from his wounds while being treated at the local hospital. Aderholt had served as officer for his community for nineteen years, the last seven as police chief. He left behind a wife and six children. His funeral was the biggest ever held in Gastonia up until that time. No one knows for sure who fired the first shots or whose gun killed the chief. Both sides accused the other of being guilty.[113]

Ella May most likely was not involved in the events that developed in Gastonia on June 7. The Workers' International Relief showed a film about the Passaic Strike at Bessemer City's Rex Theatre that night. The film was a source of controversy and was denied a showing in Gastonia. As a native union leader of Bessemer City, Ella May would have likely been there to watch the much-anticipated party propaganda film.[114]

Fear of reprisal from the Committee of 100 consumed union members for months to come following the shooting incident. The retaliation started the very night of Aderholt's shooting. Upon hearing the news of the shooting, some 2,000 locals gathered at the hospital and around the city hall. Some from the assembly decided to take revenge on the strikers. Community members were deputized on the spot. They tracked down and arrested seventy-four strikers. Male and female textile workers out on strike were thrown into the Gastonia City Jail throughout the night. At one point that evening, according to Vera Buch, the imprisoned were tear-gassed in their cells. Members of the mob destroyed the union building, tore down tents, and beat union members. Fred Beal escaped the mob that night, believing he narrowly missed a lynching. The next day Beal was arrested with his bodyguard Myers in Spartanburg, South Carolina. Fred Beal, Vera Buch, and scores of others arrested were charged with conspiracy of murder.[115]

Beal and Myers were moved from Spartanburg to Gaston County the evening of their capture. A city councilman and two police officers escorted the accused leader and his guard in a car. According to Beal, the men appeared agitated and concerned about something as they whispered to one another. The car stopped twice, once in Kings Mountain at a store for refreshments, where the escorts used the telephone. Beal overheard, "I wonder who let it out that Beal would come through tonight." Soon after, a car loaded with armed men confronted the officers and attempted to remove Beal from the car. The men guarding Beal and Myers persuaded the vigilantes to leave and the group hastily retreated to Charlotte, fearing further confrontation.[116]

On June 8 the mayor of Gastonia, Emery B. Denny, made a public statement regretfully announcing the death of Police Chief Orville Aderholt and commending his service. Mayor Denny praised the efforts of the police and special deputies in arresting the mass of union leaders and strikers. That same day the *Gastonia Gazette* made its own comments and fueled the town's anger with an article titled "Their Blood Cried Out." In this article the *Gastonia Gazette* falsely proclaimed that the NTWU leaders advertised that their motives in the community were "out to kill and destroy." While the NTWU motives were articulated as malicious and deadly, the police chief was conversely shown as a peace-loving man: "He [Aderholt] pitied rather than censured them. After he had to use desperate methods to keep his men from resorting to violence in the face of unspeakable epithets and vile abuse from this gutter scum who have come South to prey on the ignorance of a deluded people, it was the very irony of the fate that Chief Aderholt should be the victim of unjustifiable violence at the hands of these very people." The article concluded, "The blood of these officers shot down in the dark from behind cries aloud. This display of gang law must not go unavenged." A call for revenge was what this article was. Northerners soon found the Gastonia area an especially inhospitable place to be. Following the shooting incident, public transportation was monitored in attempts to intercept any outside agitators attempting to enter the area. A Massachusetts reporter from *The Nation*, Charlotte Wilder, was arrested and later released. A man from New Jersey was told to get out of town and did so promptly.[117]

The city council voted on June 10 to dismantle the strikers' tent city and headquarters in Gastonia. The land had been guarded by special deputies following the night of the police chief's death. All access to the property was denied to strikers and their leaders after the June 10 ruling and the vacant land continued to be guarded by deputies. Strikers were left homeless. In mid–June a county commissioners' meeting was held and a Gastonia City official suggested making a law prohibiting the NTWU from re-establishing a tent colony, relief store, or headquarters in Gaston County. In the end the county did nothing to prevent the NTWU from obtaining land and re-establishing itself. Eight to ten tents were re-erected on a vacant lot near the Arlington Mill. Food was distributed at the location by June 19.[118]

NTWU efforts continued in Bessemer City throughout the summer despite hostility. One of the top fundamental NTWU leaders, Paul Crouch, spoke to a crowd of hundreds on June 12. By June 15 the Bessemer City NTWU headquarters had been taken over by the city, supposedly for overdue taxes. On the same day as the takeover, a stick of

dynamite was found under the Bessemer City strikers' speaking platform by Chief of Police Bill Hoyle. The dynamite had been lit but the fuse never fully burned.[119]

By June 13, twenty-three of the roughly seventy or so individuals held from the night of the police chief's murder on June 7 were charged with crimes and the rest were released. Sixteen were charged with murder and assault and seven were held for assault solely. Those held on assault were released a few days later in lieu of $750 bail. Those held on murder charges were held without bail and included three Northern women: Vera Buch, Amy Schechter, and nineteen-year-old Sophie Melvin. The majority of the men held were local union men: seventeen-year-old J.C. Heffner, Dell Hampton, N.F. Gibson, Louis McLaughlin, Russell Knight, K.O. Byers, William McGinnis, Robert Allen, and Kelley "Red" Hendricks. The Northern male organizers imprisoned included Fred Beal, Clarence Miller, Joseph Harrison and George Carter. The group was further whittled down to the thirteen brought to trial. The ten remaining men were charged with first-degree murder and the women were charged with second-degree murder. The events taking place in Gastonia did not go unnoticed by the nation. Mother Bloor and other labor activists toured the country in an effort to gain funds and support for those imprisoned or struggling outside of prison to keep the strike alive.[120]

NTWU members made continued efforts to keep momentum going with the strike despite the fact that its key leaders were imprisoned, acts and threats of retaliation were made, and strikers continued to file back into the mills. On June 18 the largest union meeting thus far held in Bessemer City took place. Following the gathering, strikers filed into trucks and rode throughout the community in a bold public display of defiance towards their opposition and as a symbol of unity, strength, and hope to sympathizers. Bessemer City played a larger role in the strike at this point, after the imprisonment of its leaders, than it had at any other time of the conflict. Bessemer City was integral to the NTWU's unionizing efforts, and there is no doubt that Ella May was in the center of this activity.[121]

Opposite, top: **June 11, 1929, police officers and special deputies clear out and guard the strikers' tent city following Aderholt's death (courtesy Gastonia Police Department).** *Bottom:* **Headquarters and Relief Store of the Bessemer City Branch of the NTWU (courtesy Millican Pictorial History Museum).**

On June 20, Gastonia again was a central hub of NTWU activity. NTWU leaders from New York and New Jersey arrived in Gaston County, taking over where Beal, Buch, and the other imprisoned leaders had left off. It was announced that the tent colony near the Arlington Mill would grow from about ten

Top: The women charged with murder—Vera Buch, Sophie Melvin, and Amy Schechter (September 1929 *Labor Defender*, restored courtesy Millican Pictorial History Museum). *Bottom:* Nine of sixteen charged with murder. Front, left to right: Dell Hampton, Vera Buch, Sophie Melvin, Amy Schechter, J.C. Heffner. Back, left to right: K.O. Byers, N.F. Gibson, Robert Allen. Far back: Russell Knight (not pictured), Louis McLaughlin, William McGinnis, Fred Beal, Clarence Miller, Joseph Harrison, George Carter (courtesy Millican Pictorial History Museum).

tents to fifty or so. The distribution of relief food and nightly meetings were back on a regular daily schedule.[122]

A few days after the reestablishment of the NTWU's tent city, it was threatened with being dismantled. The city claimed the inhabitants of the community threatened the city's water supply due to their proximity to the city's watershed. However, the city's threat was not followed through and strikers were not evicted. Another form of intimidation manifested in the burning of a wooden cross near the tent colony on the night of Saturday, June 22. The local Ku Klux Klan later denied connection to the cross-burning incident. Due to the lack of integration of black strikers into the tent community, it was believed that the act was a scare tactic by anti-unionists and not an act of the KKK. Another blow was felt in July when the city cut off electricity to the union camp, claiming that bills were unpaid and overdue. The NTWU asserted that any unpaid bills were from prior occupants, not the strikers. The resourceful textile workers, accustomed to uncomfortable conditions, simply carried on. Lanterns were used to illuminate the camp at night.[123]

Attacks in newspapers continued. The following was printed in the *Charlotte News* about the NTWU leaders accused of Aderholt's murder: "The leaders of the National Textile Workers' Union are Communist, and are a menace to all that we hold most sacred. They believe in violence, arson, and murder. They want to destroy our institutions. They are undermining all morality, all religion. But nevertheless, they must be given a fair trial, although everyone knows that they deserve to be shot at sunrise."[124] The *Charlotte Daily News* had a different take on the trial when it printed this July 1 article: "Gaston County is desperately near the mood to try a dozen or more malcontents for murder and condemn them by what they think about God, marriage, and the Nigger— and the history of the world has shown that on the first and last of these subjects the human race, when it has tried to think, has invariably gone insane."[125] A similar perspective of the case was verbalized by the American Civil Liberties Union (ACLU) attorney Arthur Garfield Hays, who defended the NTWU members in court: "[W]hereas innocent persons are as a general rule in small danger of a conviction in the ordinary case, yet if these persons are members of a despised minority group, tried on issues involving race, color, religion, politics, or opinion, the result is almost a foregone conclusion. A thirteenth juror, Prejudice, sits in the box and hurries innocent men to their doom."[126]

The trial of those accused of Chief Aderholt's death was compared to that of the Saco-Vanzetti case. This trial had caused national controversy based on the fact that the two accused were tried on their beliefs rather than their acts. Nicola Sacco and Bartolomeo Vanzetti were charged with the murder of paymaster F.A. Parmenter and guard Alessandro Berardelli, agents in a shoe factory in Massachusetts. The accused had emigrated from Italy in 1908, were both outspoken critics of American government, and were both anarchists. The trial began in 1920 and the controversy lasted until 1927, when Sacco and Vanzetti were executed for the crimes they were convicted of. Similarities between the Aderholt and the 1925 Scopes Trial, another controversial but religion-driven trial, were made at the time as well. The Scopes Trial, also known as the Scopes Monkey Trial, challenged the law against teaching evolution in state-funded schools. The former Scopes Trial chief defense lawyer, John Randolph Neal, was hired by the ACLU for the Aderholt trial's defense and fueled comparisons between the two cases even more.[127]

Regarding the Aderholt case, Neal informed the public, "It is not communism that is on trial, but the whole system of federal justice and civil liberties, and the enlightened people of the United States will not allow another Sacco-Vanzetti case." The nation watched to see if justice would develop. The American Civil Liberties Union had been involved with the strikers early on and had given support throughout the strike. As the murder trial approached, they increased their support, and offered a reward for the arrest and conviction of those involved in the destruction of the NTWU's relief store and headquarters. During juror questioning, the defense asked potential jurors their opinions on communism in an effort to prevent another Sacco and Vanzetti case. Halfway through the jury selection, the judge put a halt to this line of questioning, stating that communism had nothing to do with the case.[128]

The trial of those accused of Aderholt's death was held in Charlotte due to the menacing mood in Gastonia. The trial began on August 26 and Judge Barnhill presided in the case after being carefully hand-picked by North Carolina Governor Max Gardner. The judge had a reputation for being fair and liberal for the times, having previously ruled that the

North Carolina Bus Company must provide appropriate service to black customers. The eyes of the nation were on North Carolina, and the governor's action in appointing a liberal judge eased the weight of the state's burden to prove its commitment to justice. The judge swiftly decided that the evidence of conspiracy would not be seen. He said, "I shall restrict the evidence to what happened on the grounds and will permit no evidence of any conspiracy except to resist the officers on the night of June 7th." Testimony "would be restricted to events bearing directly upon the shooting."[129]

Under normal circumstances in a murder case, a defendant cannot be found guilty unless it is proven that he or she personally committed the fatal act. However, there are two exceptions. If a group is involved in an illegal act and one kills, all involved can be found guilty of murder. The same is true if there is a common conspiracy to murder and it is not known who specifically fired the fatal shot. In this case no one could be found to have fired the fatal shot; therefore all were charged with murder and conspiracy to murder. The defense believed that since the strikers did not know that the police would arrive, there was no possible way there could have been a conspiracy. In addition, the defendants were not involved in an illegal act during the confrontation; therefore, the defendants could not be found guilty of murder.[130]

Jury selection began on August 28 in a hot Charlotte courtroom. Thousands had gathered outside trying to gain a spot in the stifling building. Inside, the process was a tedious one, since most of the potential jurors had already formed views on the matter. Two five-year-old twin sons of the jailer drew names from a box and then the prosecution and the defense asked questions about occupation. The prosecution wanted no mill workers or union men, and the defense shied away from those connected to mill management. Questions regarding religion, property ownership, and preconceived ideas of guilt or innocence were also asked.[131]

The selection was rather dull until potential juror J.G. Campbell was questioned. He was a Charlotte newspaper vendor who answered inquiries in a quite unconventional manner. When Campbell was asked about his religious denomination his reply was that he was Presbyterian because he was "afraid of water anyhow." He was described as "a fiery little man, who showed great relish in telling the court and counsel that

Chief Aderholt murder trial (courtesy Millican Pictorial History Museum).

nobody ever made up his mind for him." Campbell had the courtroom roaring in laughter with his quirky remarks, which Judge Barnhill eventually, but with some difficulty, put a stop to. Neither side rejected him, assuming that the other would. On September 4 the entire jury was finally selected after viewing over 600 potential jurors. According to reporters, the defense made out better than the prosecution. The jurors included two textile workers, two union members, and four small farmers.[132]

The prosecution consisted of sixteen lawyers, all financed by mill owners and mill supporters. Bulwinkle, the Manville-Jencke's attorney and the proposed leader of the Committee of 100, was part of the team. The quantity of lawyers involved in this case was unparalleled up to this point in North Carolina history, perhaps labor history. The prosecution was led by John Carpenter and Clyde Hoey, the brother-in-law of the North Carolina governor. Hoey typified the old Southern gentleman in both demeanor and appearance, down to his long flowing gray hair and signature boutonniere neatly placed in his lapel. Hoey was considered one of the best lawyers in the state. He later became governor of North Carolina himself and then a United States senator. When NTWU organizers took the stand, Hoey focused on their ideological beliefs. He questioned the Northern leaders in the following manner: "Do you believe in God?" "No." "Then you don't regard the oath you have taken on the Bible?"[133]

Nell Battle Lewis, a writer who covered the events that year, was present during the Aderholt trial and wrote the following regarding the mode of questioning at the habeas corpus hearing in Charlotte: "There are serious indications that this trial will be much more than a trial for homicide. Not only the history of the strike, but the attitude of the prosecution exhibited at the habeas corpus hearing suggest that it may very likely turn into a heresy trial."[134]

Soon after the trial began, a surprise prop was brought into the courtroom by the prosecution, purportedly for the purpose of discussing medical evidence regarding Aderholt's wounds. A life-sized wax replica of the police chief was wheeled into the courtroom fully clothed in the very bloody suit he wore the night he was shot, topped with his trademark ten-gallon hat. Much commotion resulted in the courtroom; reporters laughed, angry rejection was heard from the defense, and Aderholt's widow yelled "a half-stifled wail." The prosecution explained that the display was needed in order for the jury to understand Aderholt's wounds. Barnhill did not oblige the prosecution and ordered it removed from the courtroom. The prosecution had financed the $1,000 construction of the replica and built it in the courtroom's basement. Reported sightings of Aderholt's ghost had been made in the courthouse, and this was the likely source. The prosecution had copied a scene from a popular movie of the time, *The Trial of Mary Dugan*. In a key point in the movie a concealed wax dummy was brought into the courtroom and uncovered, to the gasps of all present. At the sight of the wax effigy, the witness on the stand was shocked into revealing critical evidence. However, no dramatic revelations were made at the sight of the police chief's effigy. After the replica was removed, the trial continued.[135]

The police officers said that union guard McGinnis fired the first shots. They testified that Beal and Buch urged the guards to shoot, yelling, "Shoot him," and "Do your duty, guards." Testimony indicated that Aderholt had been shot walking away, after dissuading Roach from entering the union camp. Upon cross-examination, the officers admitted to having no search warrant, that there was no disturbance when they arrived, and that no one knew who fired the final shot that killed Aderholt. Roach even stated, "I don't think any man could say who shot whom. They were just shooting, that is all."[136]

September 1929, juryman Campbell went insane (courtesy Millican Pictorial History Museum).

On September 9, three weeks into the trial, a mistrial was called. Campbell, the juror who was so entertaining during the jury selection, had a mental breakdown, apparently triggered by the image of the effigy of Aderholt. His breakdown was so intense that he had to be isolated in a jail cell where he insanely yelled that he wanted to be shot, released, and buried. Later interviews with some of the jurors indicated that at least five of them believed that the defendants were innocent and that it was a case of self-defense. One defense lawyer stated, "The defendants would no doubt have been cleared by this jury."[137]

The community was outraged about the mistrial and feared that there would never be retribution for their police chief's murder. The day of the mistrial, violence towards NTWU members erupted in union-ized communities and lasted for days. In Gastonia, 300 to 500 angry men raided the tent colony, assaulted union members, and destroyed copies of the *Daily Worker* and other union literature at the headquarters. A cavalcade of 100 horn-blowing cars led by a motorcycle policeman traversed Gastonia streets yelling calls such as "We're all 100% Americans and anybody that don't like it can go back to Russia.... Long live 100% Americanism!" Writer Nell Battle Lewis said, "Thousands viewed the spectacle as the cars, with drivers dashing by red signal traffic lights, blowing horns, yelling and banging sides of machines, roared through town."

Similar sights were seen in Bessemer City, where the headquarters there was infiltrated. At one point in the evening a fleet of reportedly 105 automobiles surrounded a boarding home that held organizers and strikers. While entering the house the mob sang "Praise God from Whom All Blessings Flow," then proceeded to threaten the three union members present. One member, a British-born man, was given an American flag and told to denounce the union.[138]

In Charlotte a mob surrounded the jail that held some of the defendants from the Aderholt case. They demonstrated and threatened the imprisoned union members. The mob went to the home of one of the lawyers of the defense and found that he was not home. The angry men then proceeded to the ILD office and kidnapped three strike leaders— Lell, Saylors, and Wells—and took them into the wilderness in Cabarrus County. They were flogged, and one fell unconscious. During the flogging the perpetrators heard men in the woods approaching. Fearing the law, they fled, leaving the beaten men behind in the woods. It turned out the men weren't officers of the law but rather opossum hunters. The next day fourteen men were arrested and charged for this incident, but all were eventually acquitted.[139]

Ella May boldly spoke out against the injustices her friends, coworkers, and she herself endured. She encouraged her fellow workers to do something about it. Even in the face of violence Ella did not relent. The following are words Ella spoke at a meeting on September 11 in reaction to the wave of violent acts committed by the Committee of 100. Little did she know that she soon would be their next and most violently attacked victim. After Ella's death this speech was found among her possessions, written on the backs of union leaflets. The message was later published in an issue of the NTWU's *Labor Defender*.

> I think the mill owners see they cannot send our leaders and our other boys to the chair, and last night they made a raid on the headquarters in Gastonia and also in Bessemer City. I think they thought it would scare all the workers down here and they would quit the union.

But what butts them is that it only makes us stronger. If I had not already belonged I would join now. But I joined when it first came down here and I am not ashamed of it. I can't speak to do any good, but I try. And I do anything else I can to benefit myself and other workers. And if every one of the workers could see it as I can, we would have no trouble winning, and I hope whoever was in that mob of thugs last night will have to suffer. But we can't look for the law to punish their own bunch.

We must still stick out for our rights. That is what will whip the mill owners, and they will see we are going to win. I want every worker to stick together, and if we do, we are sure to win, and if we don't stand up for our rights and we hang on the bosses, we are fighting ourselves and fighting our children and against our freedom for the working class.

Down here in the South we have never had any freedom since I can remember and I am 29 years old and I have got five children of my own and I want them to have something to want to live for and not have to slave all their days away for nothing like I have had to do.

When I came out on strike I was only making eight or nine dollars a week and working 11 hours at night. I mean I worked. I did not stop from the time I went in till I came out and I want to say ever since I came out of the American Mill on the night of the strike, I have been working for the union and I am also doing all I can for the ILD and I will continue until it is through.

If you are a worker we want you, for we are going to have a union in spite of what the boss says.[140]

The strikers attempted to regroup from the assaults. A new tent colony was erected in South Gastonia and NTWU members made plans to meet there on the evening of Saturday, September 14. The *Gastonia Gazette* gained wind of the upcoming union gathering and warned in its September 11 article: "They have been warned to stay away. If they persist in coming, they do so at their own risk. That is the word from the good people of that community who have been law abiding about as long as they can stand it."[141]

On September 12 a group of men raided the NTWU headquarters in Charlotte and arrested seven union men and charged them with "insurrection to overthrow the state of North Carolina." They were placed in a jail in Charlotte but were released a week later due to lack of evidence. It was presumed by NTWU leaders that the arrests were made in order to keep the members from attending the planned September 14 meeting. The court later denounced the Charlotte sheriff's actions in arresting the men without just cause.[142]

Gastonia's new police chief deputized American Legion men at the American Legion Hall on Saturday the 14th. The men were used to guard roads in South Gastonia in an effort to rebuff any NTWU members' attempts to reach that day's mass union meeting. If any NTWU members tried to meet, the community would be ready.[143]

5

The Music

It seemed that every development of the 1929 Loray Mill Strike was marked by the creation of a new ballad. This infusion of balladry into a strike, and indeed into all aspects of the working poor's life, was typical. During this time period ballads were companions to workers on and off the job. For Ella, ballads journeyed and transformed with her from the farm, to the logging camps, and then to the mills.

Ballads were a familiar constant in the early 1900s. The ballad was and is one of the most common forms of folk music. This historically popular oral expression is easy to compose. It consists of a set of stanzas that tell a story involving one key event. A ballad utilizes the creator's own words, usually from root experience, and is set to a previously composed familiar tune.

Famed folk singers and forgotten workers alike have made thousands of them. Balladry was an important outlet for the mill workers in North Carolina, as it was for coal miners, plantation workers, mountain men and women, and other poor folk across the South. Anyone with a story could create one and many did. Ballads were utilized as an outlet of expression and escape. Creating ballads was a cathartic experience. Whatever stirred the soul was the subject. It could be personal, such as the death of a loved one, or a shared experience, such as workplace wage cuts. Listening to ballads was a main source of entertainment and it spread a group's experiences and history.

Balladry has been shared and transformed for centuries. The ballads sung in the remote Southern woods where Ella grew up originated from ancient European balladry. Ballads traveled with the very first immigrants to America and migrated with them to the Appalachian Mountains. Passed down generation after generation, they would change yet keep their core emotion and main message. Ella was exposed to this traditional music early on in her youth. "Sweet William" and "Lord

Lovel" are a few of these early ballads she heard on the farm and later sang as a girl in the logging camps. These pieces were particularly popular, and their familiarity was widespread. Both are tragic love stories that date back hundreds of years. Love and loss were familiar ballad themes that mirrored the often tragic lives ancestors led.[1]

Another ballad Ella sang in her youth was "Little Mary Phagan." This 1913 ballad is about the murder of a young Atlanta girl in a pencil factory at the hands of the plant's superintendent. Little Mary was attacked when she approached her boss in order to get her meager $1.20 pay. This incident caused national outrage. Its melody was the foundation for Ella May's most popular ballad, "Mill Mother's Lament." The plight of despair for factory children emanates from both ballads. The boss is also the instrument of persecution in both tragedies.[2]

Thousands of young parents and young children sought factory life as a financial refuge, but once employed, many felt bound by it. Rather than being freed by additional wages, the young felt imprisoned by their dependence upon them. Wages seemed never enough to move a family forward. Often a pay slip did not provide enough for immediate needs. Some textile workers were forever in debt at the company store. As a result, textile workers felt dependent upon the boss, at his mercy, and many felt at his wrath. Meanwhile, factory hands felt their youth slipping away under hours of toil within thick brick walls. The end of their lives was their only foreseeable escape.

During bouts of helplessness and despair, ballads would rise up. They would ease the suffering, release the pain. Thoughts of hope would replace the depression. Ballads were an outlet for expressing both the working class's hardship and hope. When workers started to organize, their feelings and experiences tied to this cause were reflected in their ballads. Ballads truly reflected the direction of the times. The surviving ballads from the Loray Mill Strike (see Appendix for full lyrics of songs noted in this chapter) are the best testimony to strikers' perspectives, for they are their creation. The Loray Mill Strike was well documented due to its controversial nature, and as a result many ballads survived. However, they are a mere fraction of the number created. New ballads sprouted almost daily as events unfolded.

Ella wrote scores of songs. She jotted down lines on scraps of paper whenever she could spare a few moments. She used her own personal experiences and those of her fellow strikers for inspiration. She created songs encouraging black workers to join the union and fight alongside their white counterparts. Although unfortunately none of these survived, they were of historical importance, for it is unlikely any like them had been written before. Black workers organizing in a post–Civil War South against owners was unthinkable, since it was disturbing to whites, both rich and poor. It was an uncomfortable proposition to most blacks too, for history had taught them that speaking up often led to violent retribution. Trying to get black workers to organize was a bold, difficult task to achieve. However, like Ella's white coworkers, Ella's black coworkers were moved by her performances. They too could relate to her music and some felt moved to act.[3]

Music in general has a powerful way of affecting those listening. Unions relied heavily on music as a means of inspiration, promotion, and camaraderie. Song was a valuable tool of communication, especially among the highly illiterate workforce. Those who created music and performed it were instantaneously seen as leaders of the cause. On the flip side, balladeers were easily identifiable targets for anti-union attack. Ella's public presence as the best-known balladeer of the Loray Mill Strike made her a target.

Mill management and its supporters no doubt felt threatened by the power that ballads had in unifying the workers. Unlike union tents, they could not be ripped down; unlike union headquarters, they could not be torn apart; and unlike the Northern leaders, they could not be run out of town. These symbols of worker unity were held in the hearts and minds of the workers themselves and carried by their voices to each other. Ballads reflected their communal experiences, their communal beliefs, and their communal culture. It was nearly impossible to destroy these homespun ballads. Even when the creator dies, as in Ella's case, ballads live on, transformed again and again by following performers.

The NTWU leaders tried to persuade the strikers to sing pre-existing songs written by Northern union organizers such as "Solidarity Forever," calling out for militancy and justice. This was traditionally how Northern unions spread their message, and the songs were dictated by those on the highest chain of the organization. The leaders pushed these songs in the early stages of the strike with moderate success. However, they never truly reverberated with the Southern textile strikers; the words and feelings were foreign, unauthentic.[4]

Ralph Chaplin, a famous artist and organizer for the Industrial Workers of the World, wrote "Solidarity Forever" in 1915 after working with striking coal miners in West Virginia. It is one of the most popular union songs of all time and has been called "the anthem of the American labor movement." Chaplin said, "I wanted a song to be full of revolutionary fervor and to have a chorus that was singing and defiant." Albert Weisbord adapted his own lyrics to this famous song and his version was the one shared with the strikers in 1929. "Solidarity Forever" is sung to the tune of "The Battle Hymn of the Republic," which is the same tune as the anti–Confederate song "John Brown's Body," which Fred Beal in his later years recognized was probably inappropriate for the South.[5]

"Solidarity Forever" emphasized the importance of the organization, the union. This did not pull at Southern textile workers' heartstrings. It didn't even mesh with the nature of many of them. Former independent farmers craved independence more than unity. While solidarity is imperative for a union's survival, it was not the message that most textile strikers immediately identified with. Union leaders were not completely unaware of the differences between workers in the South and those previously organized in the North. Weisbord's version did adjust the ballad's lines to better fit the situation at hand. He used simpler language and toned down the extreme revolutionary vibe. He focused on what the Loray strikers were doing and what they needed to continue to do to be successful.[6]

Textile workers preferred to sing their own homespun music, often embracing time tested local favorites and modifying them slightly to suit their own situation and taste. The workers also constructed brand new ballads using common mountain songs for the melodies. Their music was not sung solely for the purpose of propagating a cause. Performing their songs released their own experiences, beliefs, and hopes of the strike into the world. They had a voice. Nothing truly represents textile workers better than these songs constructed from their own words.

A favorite textile ballad at this time was "A Southern Cotton Mill Rhyme," and it typifies the Southern mill experience. Within the ballad the plight of textile workers was illuminated. Workers put all their time and effort into their work with little to show for it. Paychecks brought no riches or respect. Workers were looked down upon by their middle-class neighbors, who called them "mill trash." The only hope workers had for fair treatment and reward was the afterlife, when all would be on an equal footing. The wealthy with their fine clothes and gold watches would no longer look down upon mill workers, for "when the day of Judgment comes, they'll have to shed those things."[7]

Daisy McDonald's husband sang "A Southern Cotton Mill Rhyme" at a union meeting in Charlotte, North Carolina. He later explained that he had once worked in a textile mill in Buffalo, South Carolina. A man who wove next to him would chant the words and make them up as he went along. After McDonald's performance, this song became a particularly popular one. Its universal theme of working-man suffrage appealed to workers across the South.[8]

A young striker in 1929 named Christine Patton sang her own version of "A Southern Cotton Mill Rhyme" during the Loray Mill Strike that was later documented in the NTWU's newspaper the *Daily Worker*. The NTWU leaders promoted the replacement of the word "judgment" in the second-to-last line of the ballad with the word "revolution" during the strike. No doubt the Communist Party, which did not support religion, wanted to eliminate any reference to it in its propaganda material.[9]

Pete Seeger recorded a version of "A Southern Cotton Mill Rhyme" along with a number of other textile songs from this time period, including Ella May's "Mill Mother's Lament," on his *American Industrial Ballads* album. The Smithsonian Institute later acquired the album and other folk music in order to preserve it and make it available for future generations. Kathy Kahn, Joe Glazer, Jon Sundell, Roy Berkeley, and Anne Romaine are also artists who, following Seeger, have recorded versions of this ballad. All, including Seeger, titled their versions "Let Them Wear Their Watches Fine."[10]

Early on in the strike, "Union Boys" was created by Loray Mill striker Kermit Harden. This ballad was a call for male unification. Such ballads were needed to reignite the resolve that left the men early on in the strike when they were told by union leaders they could not carry guns on the picket line. The ballad clearly states that mill ownership, management, and the local newspaper are against them, and that the union and the union leaders are their only salvation.[11]

"We Need You Most of All," by textile worker Russell Knight, was a call for scabs to rejoin the union. It was not uncommon for strikers to return to work after the initial mass walkout for short bursts of time and then return to the picket line once they earned a little pay. There was a constant rotation of textile workers moving from working in the mill to striking on the picket line to returning to the mill again. Life was extremely difficult. Many textile workers lived moment to moment and did whatever they saw best to improve their situation at any given point in time. Rather than villainizing scabs, this song pleaded with them. "We Need You Most of All" called upon the strikebreakers' sympathy towards the workers' common plight and urged them to reunite with the strikers. It was printed in the *Daily Worker*. Many of the ballads that urged workers to join the union were printed in the Communist organization's newspaper and/or the NTWU's magazine the *Labor Defender*.[12]

Many strikers created and performed ballads throughout the strike. All the ballads that survived are valuable personal testimonies. However, there was something special about Ella May. She was seen as the main minstrel of the strike. It was her striking performances that made her ballads and her so memorable. Margaret Larkin wrote: "She had a clear, true tone in her untaught voice. She sang from the chest. Full throated, unmodulated, her voice rang out in the simple monotonous tunes…. She observed the conventions of each tune carefully. In one song she would hold the last note of the line an extra measure. In another, where the old tune called for it, each line would end with an indescribable upward lift of the note, a kind of yip, like the little yelp with which cowboys vary their riding songs."[13] Vera Buch said the following about Ella May: "She would write little ballads about the strike, set them to some well-known ballad tunes, and sing them from the platform in a rich alto voice. Her rather gaunt face would light up and soften as she sang; her hazel eyes would shine; she became for the moment beautiful. She would often speak too, urging the strikers to remain firm. She appeared to be a person of unusual intelligence who grasped every feature of the strike and could explain it in her own words."[14] Daughter Millie said the following about her mother: "She told us kids that people don't want to hear you get up and talk, they won't listen. My mama would take a guitar and pick it up and start singing."

Only a handful of Ella May's songs survived, but many more were created, perhaps dozens more. The few that did survive provide rare, valuable insights into the mind of a talented, insightful, intelligent, strong-willed mill mother trying to survive. If it weren't for the strike and Ella's music, her voice would have never been heard by the world. Her socioeconomic status would have never allowed it. Vera Buch explained: "Genius is no respect of social rank, and many exceptional minds are submerged in the multitude of those for whom the denial of opportunity is complete and daily toil exhausts all energies of body and intellect. Ella Mae Wiggins was such a one until the strike gave her the opportunity briefly to blossom out, to be a human being, to devote her talents to the struggle for a union."[15]

NTWU members in the South found themselves without leaders for a short period of time following Chief Aderholt's shooting. This leaderless stretch while organizers were in jail, and before reinforcements had yet to arrive from the North, was a rather inactive time for Gastonia, but a very active time in Bessemer City. Ella May was in the thick of things, being one of Bessemer City's most involved local leaders. She tried to rally fellow strikers with "Chief Aderholt" and other songs.

The music for "Chief Aderholt" is from the 1925 ballad "Floyd Collins." "Floyd Collins" was a well-known tragic ballad about a Kentucky man who was trapped in a cave, eventually dying there. Goodhearted Floyd, like Aderholt, ignored risk to venture into a dangerous place. In "Floyd Collins" the rescue workers were the heroes, toiling day in and day out in order to try to rescue the doomed man. Ella May played upon the strikers' understanding of the revered effort put forth to free Floyd Collins and put the strikers' struggle to stand behind their leaders in the same light. The call in the original ballad was "We'll never, no, we'll never let Floyd Collin die." And Ella's call to action was "We'll never, no, we'll never let our leaders die."[16]

Writer Margaret Larkin wrote, "Better than a hundred speeches, this song ... recalls to the people that fatal night, the reign of terror that followed it, the struggle and the ultimate triumph of the union." Ella May sang this song to a crowd of 3,000 on the night prior to the Aderholt trial. It speaks of the injustice of the authorities' interference and how the workers' rights had been ignored. Yet in the same way Chief Aderholt met his doom, Manville-Jenckes too would fall, if only the workers stand together.[17]

Of course such a far-reaching cry for union support immediately following the shooting death of the well-respected Police Chief Orville Aderholt resulted in pushback. A parody of "Chief Aderholt" was sent to the editor of the *Gastonia Gazette*. A Bessemer City worker who worked at the American Mill No. 2, the same mill as Ella May, voiced a view of the NTWU opposed to that of Ella May. The ballad warned workers not to join the union "for it is just a little Hell."[18]

Countless strikers and leaders were imprisoned during the striking period. The International Labor Defense (ILD) was created out of necessity to defend these NTWU members in court. Lawyers and propaganda material were provided for by this organization. Both the law and public opinion were stacked against those charged. The ILD was the only hope imprisoned strikers and leaders had for freedom. Those depending on the ILD backed it wholeheartedly. Ella wrote "ILD Song" in order to muster support for her fellow imprisoned union members. The music for Ella May's "ILD Song" is taken from a common hymn, "Weary Pilgrim on Life's Pathway." Whereas the original hymn calls upon the weary to lay their burden upon the Lord, Ella's ballad encourages heavy-hearted textile workers to look towards the ILD to save them from their suffering. Furthermore, the chorus reiterated that if workers joined together and supported the ILD, then the ILD would bring them victory.[19]

Ella modeled her "All Around the Jailhouse" on "All Around the Watertank, Waiting for a Train." The originating ballad describes a hobo thousands of miles from home who is deeply pained. Ella's song, on the other hand, is about a union leader only a mile away from those who love her whose "heart is full of joy" and is "just a-waiting for a strike." According to Vera, Ella wrote this song about Vera's stay in prison while she was charged with the murder of Police Chief Aderholt. Vera received ballads and letters from Ella on and off throughout her stay in jail. The two were close friends and they had mutual respect and admiration for one another. Both women demonstrated gritty determination in the face of adversity and that feeling rings true in this ballad of defiance and hope. Woody Guthrie wrote the following about this ballad: "Shore do wish the hole world was a-ringin' down songs like this—beats me why they ain't."[20]

Ella May's ballad "Two Little Strikers" followed the storyline of the originating ballad "Two Little Children," about two orphaned children.

Key word changes, such as "children" to "striker" and "church door" to "union hall door," reinvented this ballad to suit the situation at hand. "Two Little Strikers" focused on the plight of strikers' children. Attacks on strikers resulted in sad, confused, lonely, parentless children. There were reports of children being orphaned throughout the strike. Parents were imprisoned, hospitalized, and in Ella May's case, murdered. Vera Buch recalled a mother who was put in the Gastonia City Jail: "One woman came in weeping. They had taken her husband also, leaving three little ones crying at home."[21]

Ella May's ballad "The Big Fat Boss and the Workers," like many striking ballads, begins with the desperate plight of the workers and ends with the hope of a better future. This particular song also contrasts the life of the greedy boss to that of his suffering workers. The workers shiver from the cold in their own homes, walk to work, and are starved, whereas the boss man "rides in a big fine car" and sleeps in "a big fine bed and dreams of his silver and gold." Strength to fight the boss and therefore hope is acquired through the union, Fred Beal, and the ILD.[22]

"Mill Mother's Lament" is the most famous of Ella's songs and was the most requested during union meetings. It was sung at her funeral by a fellow woman striker as Ella's orphaned children looked down at their mother's casket being lowered to her grave. Ella projected her own personal working mother's experience in this ballad, and in turn the plight of all working mothers was illuminated. Women were able to earn a wage in mills, which was not possible elsewhere in the workforce. However, the wages women made were very low, far less than those paid to men. Alone they were inadequate to support a family. In addition, mothers left their children at home without anyone to tend to them. Little chance of economic improvement was realized by the mothers' sacrifice. Survival was the best textile working mothers could hope for their children, and at that mothers like Ella sometimes failed. The hopelessness felt by mothers and the suffering of their children rang true for many, and this was the strikers' most moving call to unite.

> We leave our homes in the morning,
> We kiss our children good bye,
> While we slave for the bosses,
> Our children scream and cry.

This ten-year-old takes a moment to look out upon the freedom outside that she is missing. This Lincolnton, North Carolina, 1908 photo is one of many taken by Lewis Hine on his 1908–1912 journey to document child labor (photograph by Lewis Hine, National Archives).

And when we draw our money,
Our grocery bills to pay,
Not a cent to spend for clothing,
Not a cent to lay away.

And on that very evening,
Our little son will say,
"I need some shoes, Mother,
"And so does sister May."

How it grieves the heart of a mother,
You everyone must know,
But we can't buy for our children,
Our wages are too low.

It is for our little children,
That seem to us so dear,
But for us nor them, dear workers,
The bosses do not care.

But understand, all workers,
Our union they do fear,
Let's stand together, workers,
And have a union here.[23]

For Ella May and the other women strikers, their future was on the line, and more importantly, so was their children's. For this they struck and for this they sang. Women were more visible and vocal in the Loray Mill Strike than their male counterparts, both in the streets and on the stage. Some like Ella became regular staples singing at union events. Over time Ella May became famous far beyond Gaston County for her ballads. She, like other famous women balladeers such as Aunt Molly Jackson, Sarah Gunning, and Florence Reece, was noticed during her time because of her music and performance. They are remembered today because of their boldness in speaking their minds. They were seen as integral contributors to the cause of organizing. To the women's rights movement, the strong, vocal women balladeers of the South were seen as role models and champions of social justice. The women were admired for their courage to call out publicly for change. They did this during their performances and their revelations were later highlighted in the media.

Like working mothers, young workers too felt desperate for a better future. Eleven-year-old Odell Corley wrote "Up in Old Loray." The verses are sung to the tune of the popular song "On Top of Old Smokey." Odell was paid half as much as a regular hand because like many child workers she was considered only a spare. Spares were temporary workers used on a need-to-need basis. Many of these spares, although officially defined as temporary, were used quite regularly, but their pay did not mirror that reality. Young workers told employers that they were of age when in fact they were not. Employers unquestionably were aware children younger than sixteen were on their payroll but simply looked the other way. Odell, not even yet a teenager, realized how despairing her situation was, how she was being robbed of her pay and of her youth. Odell sang, "Up in old Loray, all covered with lint, That's where our shoulders, was crippled and bent."[24]

Despite the times, neither Odell's young age nor her sex impeded her ability or willingness to articulate her own bold perceptions of the strike. Odell Corley's "Let Me Sleep in Your Tent Tonight, Beal" speaks of two major events that occurred during the Loray Mill Strike, i.e., the destruction of the union hall and relief store and the eviction of mill workers from their homes. Odell laid blame for the destruction of union property on the strikebreakers. She points out that when these same scabs are thrown out of their homes, NTWU leaders will stand by them and be their only hope. The ballad is a call to scabs' morality, urging them to do what is right, stand by those who will stand up for them. She urges the scabs to rejoin the union.[25]

"Come on You Scabs If You Want to Hear" by Odell Corley is set to the tune of "Casey Jones." "Casey Jones" is about a railroad engineer who was killed when his locomotive collided with a stopped freight train. "Casey Jones" was written by a black railroad employee named Wallace Saunders in 1900. The first two stanzas of "Come On You Scabs if You Want to Hear" mimic closely Saunders' original form. Several specific details are mentioned in this ballad, like how striker Violet Jones' boss unsuccessfully tried to bribe her to return to work. Her husband Troy Jones was a loyal mill employee who tried desperately with management's help to turn his wife's affections back towards him and the mill. "Come On You Scabs if You Want to Hear" also mentions how a man was caught trying to shoot Beal with a pistol. This occurred during a union meeting. The man was removed and leaders carried on. Odell predicted in this ballad that no bribery or brutality would turn the union and so cruel Manville-Jenckens would ultimately fail. Odell did concede that money would turn some textile workers' allegiances—scabs that is—in her "I Bought a Scab for Fifty Cents." This short ballad illuminated the fact that it was the billfold that pulled the majority of the scabs' alliances, more so than any particular philosophies or ideologies.[26]

Daisy McDonald was another female balladeer who worked hard to rally union support. Daisy McDonald used the melody of "The Wreck of the Old '97" for both of her surviving ballads. The originating ballad was about a speeding train that, while trying to make up lost time, jumped the tracks in a bend, crashing to the bottom of a ravine, killing the crew. The last stanza of "The Wreck of the Old '97" calls for wives to be kind to their husbands. In both of Daisy's songs "The Speakers Didn't Mind" and "On a Summer Eve" the last stanzas call for workers to organize and support their leaders.[27]

Daisy, like Ella, left the mountains young and newly married to find a better life promised in the mills. Both Daisy and Ella were sole supporters of their families and despite their hopes and labor, failed to support their loved ones adequately. Daisy struggled to feed nine people in her household on her $12.90 weekly check. She played a supporting role in the NTWU's organizing efforts and inspired strikers from time to time with her narrative ballads. Daisy stepped up her role while leaders were imprisoned and tried to rally strikers. This was yet another similarity to Ella May, for she did the same. In "The Speakers Didn't Mind," Daisy tells the story about the night of Aderholt's shooting through the eyes of the strikers. The injustice strikers faced throughout that day is highlighted; the taunting and egg-throwing during a speech; police arriving at the tent colony agitated; the mob attacking and arresting union men. At the end of the ballad she speaks of the hope that still lies with their union leaders. Daisy pleads with strikers to be good to their leaders and themselves by organizing.[28]

"On a Summer Eve" was created by Daisy MacDonald at the end of the Loray Mill Strike, in an attempt to keep momentum moving and sentiments alive, against mill owners and for the union. It was published in the November 1929 edition of the NTWU's *Labor Defender*. "On a Summer Eve" calls out to workers to support the ILD, for the ILD will save their leaders. It also points out that Ella May died trying to do this very thing. The ballad incorrectly states that the ILD then took care of Ella's orphaned children. However, the ILD would have done this if they had been allowed to do so. The overall sentiment of "On a Summer Eve" is that strikers need to take care of their leaders because their leaders will take care of them.[29]

"On a Summer Eve" ends, "When they were put in the dirty cell, In the Gastonia jail we all know well." Evidently Daisy was one of the female strikers who spent some time in prison. Ella too was incarcerated at some point in the strike, as many were. Vera said the time the women spent in prison was passed by talking and singing ballads. She said that all the local women knew a number of them. Ella May and the others sang "Barbara Allen" and "Red River Valley." Vera also recalled singing "Birmingham" and "All Around the Water Tank" while passing time in prison. These familiar ballads spoke of love, cruelty, and remorse. To textile workers stuffed in a cramped, hard prison, singing ballads such as these would likely bring mixed feelings. Remorseful feelings would

arise from the ballads composition and the women's current predicament. However, warm, secure, comforting memories would also be
reawakened, for these were the ballads from their childhood.[30]

Ella's story and murder moved a number of balladeers to write
about her. Daisy MacDonald was one of the first with "On a Summer
Eve" in 1929. Decades later, Malvina Reynolds created "Ballad of Ella
Mae" [sic] in 1955. Malvina's inspiration was an article she read about
Ella May. Some of her lyrics are direct quotes of Ella's that were spoken
at a union gathering shortly before her murder. Notes of Ella's speeches
were found with her belongings following her death and then later
printed in various magazines and newspapers. "Ballad of Ella Mae"
restates Ella's reason for joining the union: "It's the one best thing that
I ever could do, for those babies of mine." Malvina places the blame for
Ella's children's deaths and Ella's murder both on the same source: the
company. The ballad ends on an optimistic note, that Ella May's voice
and mission continue.[31]

Gary Green wrote "The Murder of Ella May Wiggins Song" in 1977.
This song, written nearly fifty years after Ella May's death, is testimony
to the power and longevity of Ella May's story. Green discovered Ella
May's music as a teenager while playing music found in Alan Lomax,
Woody Guthrie, and Pete Seeger's book *Hard Hitting Songs for Hard-Hit
People*. Within the book Ella May is identified as one of the best song-
writers in the country.[32]

Green also discovered Gastonia in his youth. He moved there with
his parents while a sophomore in high school. Later, in college at the
University of Tennessee, Green became close to Pete Seeger and Woody
Guthrie's widow. Both encouraged him to write about what moved him.
Green moved back to Gastonia after graduating college and took a job
as a police reporter for the *Gastonia Gazette*. While employed with the
Gazette, Green worked with a man with close ties to the strike who har-
bored deep resentment towards those in the forefront of the strike,
including Ella May. This man was a Gastonia police major by the name
of Ferguson. His father worked for the same department a generation
prior and had been an eyewitness to many of the events that unfolded
in 1929. The elder Officer Ferguson escorted Police Chief Aderholt to
the tent city the night he was fatally injured. Ferguson too was shot
during the exchange of fire that night. Green was sympathetic to his

contemporary's feelings and it sparked an interest in him to dive deeper. What Green found in newspaper articles and books was a one-sided, often "red-baiting, anti-union, and pretty brutal" account. He gained a greater sympathy for the strikers themselves. At around this time Green was asked by a public television station in Charlotte to record some of Ella May's songs in a documentary about the Loray Mill Strike, which he did. All these experiences congregated together to inspire "The Murder of Ella May Wiggins Song."[33]

There are a few misconceptions in this ballad, such as her working at Loray instead of the American Mill No. 2, which is a common mistake. Ella did give birth to nine children, but they were not all living at the same time. Also she was not shot in the brain, but through the heart. However, the chorus accurately and insightfully portrays Ella May's situation. Some accept their fate, others seek better fortunes elsewhere, yet some remain firm, demanding better where they stand. This is exactly what Ella May did. Unfortunately for Ella, that day she stood firm in the back of a pickup truck, someone took her life for her doing so.[34]

Music acts as a time capsule. The ballads are the most personal type of preservation of history that we have. No documentation of the past is more direct and raw. Experiences captured in ballads are from the poor, not the rich. These intimate insights into the poor man's experiences are often the only records we have. Without this oral tradition much would be lost, for historically it is those with money and power whose stories and perspectives are told. The love of music, the necessity to create and share it, has kept the memories of our poor ancestors alive.

The ballads are now celebrated forms of music memorializing important events, like the Loray Mill Strike. The developments of a strike are recorded as the ballads are created. Even more than that, the workers' emotions tied to these events are brought to life for all to feel. The ballads are indeed powerful testimony and have been treated by historians as such. The frequency and significance of singing religious hymns during the striking period, however, are often overlooked and underappreciated. Religious music was utilized by both sides of the strike as means of validating their prospective causes. This is especially important to note during the Loray Mill Strike, because it was seen as being organized by godless communists.

Religious music sung by mill supporters emphasized that God was on their side, the side of God-fearing Americans. The opposing philosophies of the top fundamentalist Communist Party officials with those of the conservative Southern community regarding religion was often highlighted and emphasized in local newspapers throughout the strike. Headlines of Good vs. Evil, Right vs. Wrong, God vs. Satan, Friend vs. Invader, and American vs. Communist were emotional calls of morality to the people of Gaston County. The community was inundated with messages that everything they held to be sacred, just, and true was under attack. The NTWU-led Loray Mill Strike was seen as a real threat to the American way of life. The NTWU would indeed have been a real threat to religion in the United States if those pulling the strings in the Communist Party from the Soviet Union had their way. Leaders of the strike fed the fire when they proclaimed their communistic, agnostic views, and their drive to spread those views throughout the South.

Members of the community put it upon themselves to remove the cancerous threat. Any means deemed necessary, including aggressive and destructive ones, were taken. Retaliation, a call for religious unity in rebuking and ousting the ungodly presence, was vocalized in the singing of hymns during acts of violence against NTWU leaders, strikers, and sympathizers. "Praise God from Whom All Blessings Flow," for example, was sung by the crowd that pillaged the union headquarters and relief store. There were no crimes committed against those associated with the NTWU in the minds of those who saw the NTWU as evil. Seen through the eyes of religious purity, acts of violence were justified. This is nothing new in history.

However, textile workers who relied heavily on faith also sang religious hymns, believing in their hearts that God understood their suffering. Textile workers in Gaston County, including those who became members of the NTWU, viewed themselves as Christian and religious. It is true that many didn't attend church on a regular basis. Some even held disdain towards ministers, believing they were financed by the bosses and were therefore partial toward them, which they were. However, religious unity was demonstrated throughout the Loray Mill Strike by strikers. Many a "praise-the-Lord" was heard after speeches and songs. Religious rhetoric and hymns flowed spontaneously from strikers at union gatherings. Following favorites such as Ella May's "Mill Mother's Lament," crowds would burst into hymn. Poor Southern workers grew up singing religious music. Like singing ballads, singing hymns was part of their culture. Unquestionably, nothing revealed the beliefs of textile workers more than the music they chose to sing.[35]

6

The Shooting

On September 14, 1929, Ella and about twenty other Bessemer City union members prepared to travel seven miles east to South Gastonia. The NTWU had planned a rally and expected a crowd of some 2,000 individuals, but the Bessemer City group did not realize it had been canceled due to fear of retribution from the Committee of 100. The organizers were afraid of the hostile threats mill sympathizers recently made towards them both, personally and publicly.[1]

In the summer of 1929, regional newspapers reiterated and fueled the local community's frustration with the NTWU. On September 14, the *Gastonia Gazette* clearly defined villains and heroines of the community in reprints, new articles, ads, and editorials that filled the paper. The *Asheville Citizen* said that the "patience of the people of Gastonia had been terribly taxed by the Communist agitators." The *Spartanburg Herald* described the Gaston community as "harassed, overtaxed and indignant." The *Gastonia Gazette* reinforced its own position of sympathy towards its citizens' suffering and community's upheaval due to meddling outside agitators.

> The *Spartanburg Herald*, our good friend and neighbor across the line, sees the communist situation in Gastonia as we do. "Ordinarily of course, legal means of procedure must be sought. But is the law going to do to get rid of these communists? These extreme radicals, mind you, denounce every reverend belief of the American people and would abolish religion, our form of family life, our political constitution and our established order of property rights. We said the other day and we say again that if these communists have any sense they will get out and stay out.... It [the mob] broke into communist quarters and destroyed communist literature, but nothing else. It took three red agitators off and beat up one of them. Absolutely illegal and wrong, yet— for a mob—restrained action was this. Now there are mobs and other mobs. American history records mobs that have been revered. You have heard of the Boston Tea Party and similar action in Charleston. Not in our section but in the North, mobs that defied the Fugitive Slave Act and set fugitive

slaves free from their captors and were never touched by officials. The original Ku Klux Klan was absolutely illegal in its activities, yet at the time it was born the people thanked God for its coming.... We cite those historical cases merely to show how a harassed, over-taxed and indignant people, seeing no way by legal methods to get rid of a burden or an offense to their moral sense, will break through and take direct action as a means to obtain relief."

Such justifications as these were no doubt in the minds of those who confronted Ella and her fellow strikers that day. The same paper also printed a full-page ad titled "LORAY—THE MILL WITH A PURPOSE." The ad touted the amenities of the Loray Mill, including its splendid climate, new equipment, and the personal touch of the boss. It declared in bold print: "LORAY THE MILL WHERE THE BOSS IS YOUR FRIEND."[2]

News of the September 14 mass meeting of NTWU members caused a vocal uproar from a community on edge. Threats against NTWU affiliates and calls to a community to unite in defense were made. The kidnappings and beating on September 10 were attempts to thwart those planning to attend the meeting. Individuals in that mob threatened their victims that they would personally stop the upcoming mass meeting if it proceeded as planned. The *Gastonia Gazette* voiced that the communists should stay out of Gastonia. It warned that trouble would erupt if they followed through with their meeting. The day following the September 10 mob violence, NTWU leader Bill Dunne stood resolute about gathering on September 14: "We will have the meeting at South Gastonia where it had originally been planned to have it; it will be at 3 o'clock and will be held just as if there were no interruption." He changed his mind as that date neared and threats escalated.[3]

H.W. Branch, the only Committee of 100 member to later talk openly in court about the group's activities and motives, disclosed that on September 14 the Loray Mill's intelligence officer gave him a pistol, twenty rounds of ammunition, directions to "do everything necessary to break up the union meeting," and an excuse note to give his foreman. The note stated that he was out "serving a good cause." According to Branch, many Committee of 100 members were prepared in a similar fashion by agents of the mill.[4]

Twenty-five anti-union armed men, many deputized that day at the

American Legion Hall, combed the area where the meeting was to be held. Roads to the site were blockaded with cars. Only one known car made it to the proposed rally and its occupants were searched upon arrival. In that car were writer Mary Heaten Vorse; Liston Oak, the publicity director of the ILD; Oak's wife Margaret Oak; and the chauffer, C.M. Grier, all Northerners. After a search, Oak was charged with carrying a concealed weapon, Grier was charged with reckless driving, and both were taken to jail.[5]

Two dozen or so Bessemer City NTWU members gathered at their Bessemer City union headquarters. Ella, seven months pregnant, her lover Charlie, her brother Wesley, and the other NTWU members climbed aboard the bed of a rented pickup truck. Some men in the group wanted to bring guns for protection. This was a longstanding debate among many textile strikers and their leaders. Ella was the one who convinced the men in the truck to travel unarmed. As the truck pulled out of Bessemer City, tensions were high. Charlie overheard a young boy in the truck predict that they all were going to get shot. When the group stopped at the union headquarters in Gastonia, they were met by a large blockade of cars. When the truck stopped in front of the blockade, the driver was told to turn around, leave town, and head back to Bessemer City or he would be shot. The driver, a hired man unconnected to the NTWU, turned the truck around and started driving back to Bessemer City.[6]

The truck was followed by half a dozen cars. Just past Gambles' Crossing, an iron bridge overpass, the first car in the line, a dark black Essex, sped ahead of the truck, pulled in front of it, and suddenly stopped. The driver of the truck was unable to avoid slamming into the car. Some passengers fell to the bottom of the truckbed, others fell to the ground, and some immediately jumped out and fled through nearby cotton fields. Ella remained, standing firm on the left-hand side of the truck next to the railing near Charlie.[7]

The Essex that had been hit was pushed off the road in the collision. The four occupants of that car stumbled out, some brandishing guns. Within seconds all of the cars involved in the pursuit stopped and dozens of men stood outside with guns pointed towards the strikers. Tension filled the air and each second seemed like an agonizing eternity as strikers and anti-unionists waited for violence to erupt. Ella stood firm in the bed of the pickup truck staring down at the aggressors. Suddenly a shot broke through the silence. The bullet struck Ella, ripping through her heart and lodging in her spine. She clasped her face and yelled, "Lord a mercy, they've done shot and killed me." Charlie caught her a moment later as she collapsed. Ella May was dead, and with her death her unborn child died as well. Charlie carried her body to a nearby home and laid her out on its wooden porch.[8]

Charlie recalled the incident in his own words:

> We was on this truck goin' to the speakin' at South Gastonia, Ella May an' me an' Wes an' a good bunch o' people. We'd got about halfway down past the Arlington Mill where they got the tent colony now, then they was a lot o'cars standin' in the road an' a great crowd o' men, the Loray Mill gang, the deputies, the same ones that raided this headquarters. An' they told our driver he'd got to turn the truck back. What could we do … we didn't have no guns an' they was all armed. So the truck turned back for Bessemer City. About five o' the cars was followin's us. We'd got as fur as Gambles' Crossin', where they's an iron bridge. We'd just got over the bridge when this Essex coach run right around in front o' us an' we collided with it. So we was stopped an' it was then that Ella May was standin' at the railin' of the truck leanin' against it, an' a man come up not fifteen feet away and raised a gun an' fired at her.[9]

After that first shot that killed Ella, a series of shots followed and chaos ensued. Forty to fifty shots were fired as the occupants of the truck jumped out and fled the scene. Individuals being shot at scattered through a cotton field and nearby woods. The armed men ordered the remaining NTWU members to hold their hands up in the air. Some strikers were beaten with blackjacks. One man's arm was broken. The driver of the truck, Lingerfelt, was blamed for the accident by the mob on the scene. He was beaten with a blackjack and forced into a car by the mob. He was taken to the sheriff's office in Gastonia. When he was about to be released, American Mill superintendent Spencer arrived and ordered him to be locked up. Officers obliged.[10]

Spencer said he just happened to be passing by as the truck and car crashed and shots were fired. He later testified to seeing men in groups shooting at others fleeing. He was able to identify Lingerfelt as the cause of the accident, but unable to identify any specific individuals involved in the shooting.[11]

Ella's children believed that on September 14 their mother had been enjoying herself at a picnic. Apparently that's what Ella called her union events in order to ease her children's minds by letting them think that she was not in danger. When their mother didn't return later that evening the children walked to a black neighbor's house in order to borrow some cooking oil for supper. The man who answered the door knew what had happened to their mother and informed them. The children turned around, devastated, and began walking home. Millie recalled vividly years later following the red clay road with tears in her eyes, huddled with her siblings. A couple in a coupe rambled up the road and at the sight of the distraught children stopped and asked them what had happened. After hearing the news they offered to take in three-year-old Albert. Ella May had always told her children to stick together, so Myrtle, with the youngest (Charlotte) on her hip, kindly refused the offer. The children returned home without a mother, but together. The children were not alone for long. Charlie stated that soon after Ella's death he located the children and comforted them.[12]

The children were then brought to the local mortician's office where their mother's body was being held. Frank Sisk was the mortician who both worked and resided there. While standing in his office the children overheard Sisk receive a call from a man threatening to take their mother's body. Millie overheard the mortician tell a friend "someone might steal the kids." Ella's son remembered angry men with torches, guns, and rocks. Ella's children believed the men were after them as well as their mother's corpse. Charlie disappeared. It was Sisk who kept Ella's children as well as her body safe at his own home that evening. He called some friends to help him guard the funeral parlor and sat up all night on the front porch with a shotgun across his knees. Ella's children viewed Sisk as a hero and some of them kept in touch with him well into adulthood. The NTWU tried to gain access to the children, but were denied. Instead they cloaked the Bessemer City Union Hall and Ella May's home in black. On Ella May's home they hung a sign that read "Ella May, slaughtered by the bosses Black Hundred, martyr to the cause of organized labor."[13]

That evening Governor Max Gardner went on record, saying in regard to the shooting: "Irrespective of its source or the occasion, there is no place in North Carolina for terrorism in any form." The governor then further explained:

Ella May's children, September 1929. Left to right: Albert, Myrtle holding Charlotte, Millie, and Clyde (October 1929 *Labor Defender*, restored courtesy Millican Pictorial History Museum).

I want it distinctly understood inside and outside North Carolina that under our system of government the law is supreme. Under it every citizen is guaranteed the protection, not only of life but of civil and property rights, including the right to work unmolested, the security of property, freedom of thought and of speech and the right to peaceful assembly.

It is deeply regretted that some of our citizens have allowed themselves to become provoked to the point where they have attempted to administer punishment without due process of law. No matter how acute the provocation, constitutional government rests upon law and order.

Acts of violence will not be tolerated, and it is my purpose to employ every means within my power to detect and punish to the full limit of the law all persons who have engaged or who may hereafter engage in this form of lawlessness.

Under no circumstances can violence or terrorism be justified and in no case and in no circumstance will either be tolerated, and this irrespective of whether it emanates from within or without the State.

Any form of anarchy or any form of terroristic communism occupy alike a position without the pale of legal sanction and both will be promptly dealt with as crimes against the State of North Carolina.[14]

North Carolina newspapers soon voiced their varied interpretations and opinions over the circumstances and significance of Ella May's death. The *Gastonia Gazette* said that it regretted the death of Ella May and in no way condoned the breaking of the law, but was concerned that the communists would use the incident to their advantage. It said that the citizens of Gaston County were "playing into the hands of the communists" and in turn bringing "disgrace and shame." The paper said that Gastonia wanted law and order and deplored mob violence. The paper tainted its own sincerity by stating that the murder of Ella "shades into insignificance when compared with the lawlessness of the communists at Loray." The *Gastonia Gazette* even hinted at pointing blame towards Ella May herself for choosing the NTWU over her young children: "With all respect for the unfortunate woman who became the victim of a gunshot, the question naturally arises why was the mother of five children ranging in age from 13 months to 11 years, in the gang of Communists and Communistic sympathizers traveling from Bessemer City to Gastonia? Of course, the answer will be, she was on her way to attend the scheduled 'red' rally in South Gastonia, and that she had a legal right to be where she was. True, but where was her first duty? Was it attending a red rally, or looking after her home and family?"[15] The *Gastonia Gazette* continued: "It is true that a Communist has little time for home and that they generally hold that children are more care of the state than of the parents, but we had hardly thought that such doctrines had yet found firm hold on Americans."[16] Apparently to the 1929 *Gastonia Gazette*, textile strikers leaving their children on September 14 to organize in order to improve their pay and therefore improve their children's plight were being neglectful parents. However, there was no mention of those community members who left their children to impede the strikers, to inflict violence toward them, and how they were being bad parents.

A more liberal paper of the time, the *Asheville Citizen* had an opposing perspective of Ella's death: "There has been no more defenseless murder in North Carolina than the wanton killing of Ella May Wiggins. But she was a Communist sympathizer. So far as Gaston County is concerned, she had put herself outside the pale." The *Greensboro Daily News* stated, "As calmly and casually as if it were an every-day occurrence, Gaston County wipes its hands of the blood of Ella May Wiggins and turns to more pleasing prospects." The *New York World* predicted the outcome of an investigation of those involved in Ella's murder when it wrote, "Lynchings are investigated frequently. Convictions are rare." The newspaper also pointed out that this is no new phenomenon, nor limited to the South.[17]

The NTWU immediately propelled Ella May's status to that of a martyr. Her personal story of hardship and struggle to fight for the union was shared in the *Labor Defender*, *Daily Worker*, and other instruments of party propaganda. The NTWU tried for months to use Ella's death as a rallying call for unification and change.

So was Ella May's death a random act of violence against a NTWU striker? Did the shooter get caught up in the heat of the moment and shoot wildly and by chance strike Ella May? It is very unlikely that Ella May's murder was in any way accidental. It is more likely that the shooter knew exactly what he was doing and who he was doing it to. Ella May stood still and the shooter was only a few feet away. There were twenty other strikers around her and many of them were men, who were normally preferred targets. Ella was well recognized as the minstrel of the strike and as the organizer of black workers. Either would be cause enough for some to kill her. The latter was the more likely reason, for black hatred in the South ran deep. Vera underscored why they specifically targeted Ella May when she wrote: "[Ella May] understood immediately without

Caricature of Ella May after her murder for *Labor Defender Magazine* (cover of October 1929 *Labor Defender*, restored courtesy Millican Pictorial History Museum).

argument the value of our union principle of racial equality. What is more, she set about of her own accord to organize the black people in Bessemer City, where she lived. Even more than Communism, it was the appeal to the black people, and especially her role in their organization, that incensed mill owners and like-minded people in the South. I am certain it was as an organizer of the Negroes that Mrs. Wiggins was killed."[18]

7

Aftermath

The voice that articulated the heartbreak and hopes of textile mothers everywhere was forever silenced. On September 17, 1929, a dim rainy Tuesday morning, textile workers gathered to say farewell. Scores followed Ella's body from the funeral parlor, along wet main streets to the local cemetery. As the procession solemnly proceeded by the brick walls and wire fences of the local textile plants, workers left their stations and crowded around windows to take a moment to witness and pay tribute. At American Mill No. 1, workers boldly broke the rules, left the building and gathered at the locked gate to pay respect. Others took off work altogether to participate in the funeral. Several hundred strikers gathered at the Bessemer City Memorial Cemetery, many piling out of beat-up cars, and many others crossed cotton fields on foot to say farewell to the minstrel of their strike. Six textile workers from Bessemer City carried Ella's coffin to the red clay earth that marked her final resting spot.[1]

Ella was buried in a $10 unmarked grave. The Marble Workers Union of Vermont had sent a white monument for her gravesite, but due to concerns over possible defacement it was never used. Ella's brother Wesley placed a large rock on her resting place instead. The Reverend C.J. Black, a Bessemer City Baptist minister called by the undertaker Frank Sisk, held the service and made no reference to Ella May's involvement in the union, nor the circumstances of her death. The reverend was a known outspoken opponent of outsiders. The NTWU leadership protested the choice of the Baptist minister, but Sisk's pick won out. Sisk wanted to see to it that Ella May had a proper Christian burial. The reverend ended the service from a reading with John 14, "In my Father's house there are many mansions…"[2]

Fellow strikers spoke at Ella's funeral. One said, "You all knew our sister, Ella May. She was one of our best workers, and we'll feel her loss,

I reckon. Her death is on Manville-Jenckes and on North Carolina, too. She died for us and the union. We must go on fighting."[3]

Ella's children, dressed neatly in borrowed clean clothes, were brought forward. Onlookers snapped photographs of the children standing in front of their mother's grave with solemn, drained expressions. The children placed flowers on their mother's casket. "Mill Mother's Lament" was sung by striker Kathy Barrett as Ella May's children witnessed her coffin being lowered into her grave. Adjoining lots were reserved for all of Ella May's surviving children. Myrtle and Albert were later buried alongside their mother, while the other children were buried elsewhere.[4]

Millie remembered at the funeral overhearing representatives of the local churches arguing over who would have to take in the tattered, "unkempt" children of the controversial troublemaker. Millie recalled that no one wanted them except the NTWU. The NTWU wanted to take the children north and enter them into the Young Pioneers School in Philadelphia. The school would take care of them and teach them the philosophy of communism. Not coincidently, the children would also be readily available to do propaganda work for the NTWU.[5]

Despite the fact that no local religious group wanted to take responsibility for the children, none of them wanted children from the South to be taken by Northern communists. During the funeral it was announced by two ministers that the children would be placed in a nearby orphanage. The older children were taken to the Presbyterian Orphans Home in Barium Springs. The youngest, Charlotte, who was too young to be placed in an orphanage at that time, was handed over to her aunt and uncle, the Pickelsimers, who lived in the Loray Mill Village. In the weeks following Ella's death, NTWU representatives visited the Pickelsimers' home on a number of occasions in an effort to persuade them to allow the NTWU to pull the children from the orphanage. Their efforts were not fruitful and the NTWU never did intercede in the lives of the children. The children grew up God-fearing and Southern.[6]

Throughout the years, Ella May's older children spent time floating between the Barium Springs Orphanage and various family members' homes. Records from the orphanage indicate that Millie spent time there from September 17, 1929, through December 27, 1934. The children didn't care for institutional life, but they were fed, clothed well, given

September 17, 1929, Ella May's children at her funeral. Left to right: Clyde, Millie, Albert, Myrtle holding Charlotte (October 1929 *Labor Defender*, restored courtesy Millican Pictorial History Museum).

medical care, and educated. They did not take these basic necessities that they lacked for many years for granted. However, they did not feel loved, and this was one thing Ella May had been able to give them.[7]

Decades later I discovered "My Mother—The Queen of Hearts" written by hand on the back of a family picture of Ella May, testimony from Myrtle of the love her mother gave. The children grew up with a sense of longing for their mother. To Ella May her children had been everything, and reciprocally Ella May had been everything to her children. Affection became something foreign to Ella May's children once she died. Even more, they felt invisible. Albert was seen by his siblings as being given some special treatment due to the fact he was the youngest child at the Barium Springs Orphanage. Staff members would dress him up and gush over how cute he looked. However, Albert too felt a hardness and lack of love. He was a nervous child and was seen as too much to deal with at times. When he cried, busy staff would lock him in a closet.[8]

Top: Barium Springs Orphanage. Picture of Ella May's children in 1933. Left to right: Millie, Myrtle, Clyde, and Albert. *Bottom:* Barium Springs Orphanage. Picture of Ella May's children in 1933. Left to right: Albert, Clyde, Myrtle, and Millie (both photographs, author's collection).

Charlotte, from the age of four through nine, was sent away from relatives in Gastonia to live in an orphanage by herself in Asheville, North Carolina. She attested to harsh treatment in that facility. Charlotte was sometimes tied to a hot radiator and other times beaten for trying to go to bed early. Going to bed early was frowned down upon in orphanages during that time, for Millie too remembered having to stand in a hallway at the Barium Springs Orphanage, banned from going to bed early. Millie found a way to make herself and her siblings sneeze so the staff would think they were sick and let them go to bed. All of Ella May's children tried to escape orphanage life when they could. They preferred living in their relatives' homes in Gaston County. The children spent a significant amount of time with their Aunt and Uncle Pickelsimer, as well as with Ella's brother Wesley May and his wife Mary. Charlotte moved in permanently with her Uncle Wesley and Aunt Mary at the age of nine and developed a child/parent relationship with them. According to the older children, they were not welcomed or treated as well as Charlotte at their uncle's home. Wesley was fair to all, according to Millie, but his wife treated them harshly and beat them regularly. Millie remembered going to the well for water and fearing to return, for if she dropped any of it she would be beaten. The experiences the children had hardened them. Charlotte admitted, "There isn't one of Mom's children that you can go up to and put an arm around because they'll walk away. An arm raised to them is an arm that's going to hit you." Ella's children paid a great cost for their mother's murder and it lasted a lifetime.[9]

On the day of Ella's burial, confrontation and retaliation from both sides was feared. Additional violence was at the forefront of everyone's mind. The NTWU called for a day of protest and strike to remember Ella and what she had died for. Leaflets were distributed by NTWU organizers advertising this one-day demonstration. While union leaders planned their protest, they were concerned that anti-unionists would do the same and impede the funeral procession with their own rally. However, on the day of Ella's funeral no vocal or visible protest occurred from either side. The only united activity was found in textile workers paying homage to Ella May during her funeral procession and funeral. The next day the *Gastonia Gazette* declared its victory over communism when it printed, "In a little mud-hole to the northeast of Bessemer City Tuesday morning the first North Carolina communist movement was buried."[10]

Although violence was tempered for the day of the funeral, peace in Gaston County did not last long. On September 18, around 1 a.m., a twenty-four-year-old Kings Mountain NTWU organizer, Cleo Tesnair, was taken across the South Carolina border and severely beaten. Tesnair was forcibly removed from his home after a mob kicked down his front door. He was grabbed by the throat, thrown into a car, driven to a wooded location, and viscously clubbed. One of the attackers looked at a nearby tree and declared that it would be a good place to hang him. Tesnair was not lynched, but rather released into the woods and shot at as he fled. Tesnair's kidnapping and flogging was similar to that of organizer Ben Wells ten days prior. Wells also reported to authorities that on the night of Tesnair's abduction a carload of men attempted to kidnap him yet again, this time while he was walking in Charlotte. Wells fled and successfully found shelter at the Charlotte ILD headquarters.[11]

An NTWU speaking platform in Kings Mountain that was used every Saturday by NTWU leaders was also dynamited on the 18th. The speaking stand was located in a vacant lot and was destroyed thirty minutes before Tesnair was kidnapped. The explosion was so loud that it was said to have awoken the whole city. After this evening of aggression the NTWU suspended all union activity in Gastonia, and by September 20 there were no more public demonstrations or meetings by the organization in the area. On September 27 the leaders of the union officially announced the end of the strike. The tent colony was dismantled at this time, and organizers and strikers dispersed. The flickering hope for a better life that strikers had held onto was extinguished, violently squelched by those living within their very own community.[12]

The Charlotte NTWU office remained open, but by 1930 the party moved its focus away from textile mills in the Piedmont and towards other industries near Birmingham and Chattanooga. The majority of the NTWU's Gastonia leaders, who were so revered by the Communist Party in 1929, either voluntarily left the organization disenfranchised or were removed from their positions forcibly, blamed for the Gastonia failure. The American Federation of Labor, which was rumored to have been bracing throughout 1929 to pick up the pieces following the departure of the NTWU, never did. No doubt the AFL too feared the violence that ran out the NTWU.[13]

A number of brutal Southern textile strikes dotted the South in

1929. The Marion strike led by the AFL especially pulled national attention and spurred controversy. The violence there peaked on October 2 when the county sheriff and deputies shot tear gas into a crowd of picketers. As the strikers retreated, the sheriff and deputies opened fire on the crowd. Five men were killed and ten wounded during the incident. All shots were fired into strikers' backs. A grand jury indicted eight deputies, but all were later acquitted. The strikers' legal team in this case, as in the Aderholt trial, faced off and lost to the well-respected and renowned Clyde Hoey.[14]

From April 1929 through April 1930 the American Civil Liberties Union documented 7 deaths, 24 injuries, and 7 kidnappings in North Carolina directed towards textile strikers. No individuals were ever convicted for these crimes. In contrast, according to the ACLU every striker who was brought to trial that year was convicted of the crime for which he or she was charged. These statistics highlight how one-sided the law was at this time against those associated with unions. Ella May's murder case was no exception.[15]

Two days following Ella's murder, seven men were held and charged with the conspiracy to kill her. Four of the men accused, Fred T. Morrow, Larry R. Davis, Theodore Sims, and Troy Jones, were in the car that forced the Bessemer City strikers' truck to crash. Fred T. Morrow was the driver of the car and a Loray Mill-employed electrician. Larry R. Davis was a young Loray Mill employee. Theodore Sims was a high-level employee of the Loray Mill. Troy Jones was a Loray Mill worker whose wife Violet Jones worked closely with the NTWU, even traveling to New York to raise relief funds. Jones had charged Beal with abduction for his wife's actions. Charges were later dropped when his wife testified in court that she had gone voluntarily. He also sued the NTWU for $5,000 for alienation of his wife's affections. This too did not hold up in court. Jones previously had threatened to blow up the union headquarters and was caught with dynamite in his hand while NTWU organizer George Pershing was on the speaking stand. Despite admitting that he was there to kill Pershing, Jones was never given more than a warning for his behavior. Troy Jones's brother Loyal years later acknowledged that Troy had been a welterweight boxer and a lifelong loyal employee of the mill, and that it is reasonable that he was one of those hired by the Loray Mill to impede the organizers.[16]

Two more charged with Ella May's murder arrived on the scene in cars moments after the crash. L.M. Sossoman was another high-level employee of the Loray Mill who directed during the shooting, "Boys, don't shoot anymore." Will Lunsford was a foreman at the Loray Mill and said, "That's enough shooting, boys." Lingerfelt, the driver of the strikers' vehicle, oddly was charged as well, but later dropped from the list. Three more names were later added to the list: Jack O. Carver, W.M. Borders, and a redheaded man from west Gastonia named Horace Wheelus. Wheelus, a loom fixer for the Loray Mill, is the man whom evidence indicates is guilty for firing the shot that killed Ella May. Baugh, the Loray Mill manager, made the $1,000 bail for all of those accused of the murder.[17]

The coroner's investigation following Ella's murder led to half a dozen sessions of inquest. Seventy to seventy-five witnesses were summoned to testify regarding the incident. Everyone on the truck that afternoon testified to what they experienced. Charlie took the stand in his bloodstained clothes the day after the shooting and explained why he had been there: "I heard it discussed at Bessemer City Friday night at a union meeting that there was a-going to be a big meeting Saturday at Gastonia. We all 'lowed we'd go down there this Saturday and we-all went. I 'lowed to hear one of the organizer fellows and I reckon the rest of the twenty-two folks in that truck went for the same reason."[18] Few interviewed could or would identify specific individuals at the scene responsible for the bloodshed.

On October 23, the Ella May murder investigation findings were turned over to a grand jury of Gaston County, which determined that there was not enough evidence to indict anyone for Ella's death. Judge Sink, who presided, remarked that he was "amazed" by their findings. Outside Gastonia the rest of the country was outraged. The *News and Observer*, a newspaper located in Raleigh, was furious about the findings. In an article titled "Ella May Wiggins Doesn't Count," the paper said the following: "The blood of Ella May Wiggins is as futile as her life. Yesterday Gaston County washed its hands of her blood as easily and as indifferently as a mob of its citizens shot her down…. Her five children are in an orphanage, where they will be taught to love their State, to obey its laws, to honor justice who stands with a keen sword to protect the rights of the weak as well as the strong, the poor as well as the rich, the weary cotton mill hand as well as the mighty mill baron."[19]

The very afternoon the grand jury failed to indict the seven charged with Ella May's death, the same jury also reached the same decision concerning the four men charged with the kidnapping of three from NTWU headquarters. On October 21, two days prior, the seven NTWU members still charged with Aderholt's murder—Beal, Carter, Miller, Harrison, McGinnis, McLoughlin, and Harrison—were convicted after a three week trial and less than an hour of deliberation. Shockingly, the jury came back with the finding of guilty of murder in the first degree despite there never being any charge of such a high count. The judge rectified the draconian ruling and adjusted their sentencing to a second-degree charge. After this second trial, the defendants were sentenced to seventeen to twenty years in prison.[20]

Beal and the others convicted and awaiting appeal at the State Supreme Court received $5,000 in bail each. There are conflicting reports as to whether this bail was financed by the ACLU or the ILD. Soon after receiving the funds they all escaped to Russia, reneging on their promises to the workers to stick it out. Excited and eager to be immersed in the communist homeland, Beal absorbed himself in public relations work for the Communist Party in Russia. He canvassed the country participating in a communist propaganda campaign touting the ills of capitalism. Not long after starting this campaign, Beal's beliefs were turned upside down. He found not only the same problems as found in the United States, but far worse conditions. Young children were exploited in labor, there were beggars and slums, and in greater numbers and to a greater extreme. In addition, the government restrained freedoms such as the freedom of the press enjoyed in America. When the government censors refused to print some of Beal's material, one official said, "Where do you think you are, in America?"[21]

Beal later described the Soviet Union in his autobiography as "the grandest fraud of history." Beal returned to the United States in 1933 and in 1938 was apprehended. Beal served his sentence in North Carolina until he was paroled in 1942. Beal was released seven years early when Governor J. Melville Broughton accepted an early release request. Broughton's predecessor, Clyde Hoey, who had served as one of the leading prosecution lawyers and had helped convict Beal, had rejected an earlier plea in 1939.[22]

If there were any questions about Beal's true final feelings towards the Communist Party, they were made clear in Gastonia in 1948. He faced a judge in order to restore his American citizenship that was taken away in 1929. Beal declared in front of the judge, "I am one of the greatest foes of Communism in America. I would rather be an American prisoner than a free man in Russia." Around this time, Beal tracked down Ella's children to share with them his philosophical discoveries. He gave Ella's eldest, Myrtle, now a mother with her own brood of young children, a copy of his book *The Red Fraud: An Exposure of Stalinism*. Myrtle and the rest of her siblings, however, needed no encouragement to stay away from the disillusionment of communism. They grew up orphans, motherless and unloved, and they were taught to believe that this was because of their mother's involvement with communism. Both community members and family members alike reinforced this belief.[23]

Vera, too, felt disenfranchised by the Communist Party. After the Loray Mill Strike she was ignored and weeded from party propaganda. She never received another assignment from the NTWU, and neither did her future husband and party leader Albert Weisbord. By 1930 both had left the Communist Party. Vera kept up with Ella's children. She corresponded by mail with at least Millie well into Millie's adulthood. The vast majority of the NTWU leaders sooner or later abandoned the party. An exception was Sophie Melvin. Melvin, while critical of the party's unprepared movement into the South, viewed their overall situation as a success. She saw a battle of thousands of workers standing united under communist leadership in the South against economic and social ills, including racial inequality, as a breakthrough moment in time. The Communist Party chose to portray the Loray Mill Strike in this same light, sending Melvin to speak at various locations on their behalf long after the strike had ended.[24]

On October 26, due to outside pressure, the state began another investigation into the death of Ella May. The governor of North Carolina, Oliver Gardner, assigned a special judge and prosecutor to reopen the case. Gardner also offered $400 for information leading to a conviction in the killing of Ella May. The governor called the act "an indefensible murder." The ACLU also had an open offer of $1,000 for information that would convict those involved in Ella May's death.[25]

The state investigation found enough evidence against five of the men indicted in Ella's death to be tried on second-degree murder charges. The accused charged on January 13, 1930, were Horace Wheelus, Fred Morrow, Larry Davis, Troy Jones, and O.H. Lunsford, who were all employees of the Loray Mill. Horace Wheelus was the one responsible for firing the fatal shot, according to witnesses during the preliminary hearing. All were charged with conspiracy to murder. The Manville-Jenckes Company again furnished bail to all accused.[26]

The trial began on February 24 in Charlotte. In all, 150 witnesses were seen, about 100 for the prosecution and about 50 for the defense. Both sides had strong, experienced legal representation. Solicitor John G. Carpenter and Attorney General Dennis G. Grummitt led the prosecution. The defense attorneys were A.L. Bulwinkle, A.C. Jones, R.G. Cherry, E.R. Warren, and Plummer Stewart. This defense, like the Aderholt prosecution team, had a future North Carolina governor representing those accused, Greg Cherry. Cherry was also a former state commander of the American Legion. He reasoned that the strikers must have had guns themselves because "cats like that don't go unarmed." Cherry was clever at working witnesses. He planted a man from a nearby café in the courtroom who looked like defendant Jones. Then he asked a young striker who was beaten up on the day of the murder by Jones to point to the man who beat him. Mistakenly he pointed to the planted man. Similar tactics were used on a number of prosecution witnesses. Besides discrediting witnesses, the defense tried to discredit the victim. The defense claimed that those accused should not be found guilty because Ella May believed in communism. They argued that she gave speeches that denounced God, encouraged the overthrow of the government, and promoted racial equality. The last of these accusations is the only one with any merit.[27]

There was also an attempt made by the defense team to blame the shooting on the strikers themselves. They asserted that the bullet that pierced Ella's heart came from inside the cab of the truck. Bulwinkle passed the bullet that had lodged in Ella May's spine around to the jurymen, who examined it with a magnifying glass. He claimed that the scratches on the bullet were made when it passed through the back panel of glass in the truck. The defense cited a hole in the back glass panel as further evidence. Strikers testified that the glass was broken in the crash. The hole, which was found to be the size of two fists, was apparently made when the crutch of a crippled boy on the truck jumbled about during the impact. Members of the mob claimed that the original hole was the size of a dime and that someone enlarged it later. No one ever testified or gave any evidence to indicate that any of the strikers or the hired driver had a gun.[28]

On February 28, the trial was postponed for a few days because one of the defendants, O.H. Lunsford, had the mumps. During the lull in the trial, authorities searched a prosecution witness' property. Julius "Crip" Fowler was found to have a dozen half-gallon fruit jars of moonshine hidden outside his house. Fowler was arrested and his testimony was put into question. While a key witness' character was being discredited, the Loray Mill was working on rewarding employees for touting how great it was. During the trial break, a full-page ad appeared in the *Gastonia Gazette* offering prizes for the best articles on the topic: "Why I Enjoy Working at the Loray." Five-, ten-, and fifteen-dollar prizes were offered. Those judging included the mayor and the local Methodist minister.[29]

After a four-day recess the trial resumed. Loray employment agent Ed Spencer testified to the good character of those accused. Defendants Davis, Morrow, and Jones acknowledged their presence on the scene but claimed they were unarmed, uninvolved in any acts of aggression, and unaware that any such acts took place. These three defendants' wives testified to their husbands' innocence and stated their husbands were on their way to Lake James at Camp Jenckes to fish. A fourth man who was on this alleged fishing trip, Theodore Sims, had been on the list of those accused of Ella May's murder, but was later dropped. Sims and the three others who claimed to be en route to the camp were in the car that pulled in front of the truck. Sims claimed that he was shaken after the car accident and immediately went home. He testified that there was no mob, that he was unaware of any shooting, and that he just noticed about a dozen cars stopped on the highway due to the wreck.[30]

Amazingly, a number of witnesses for the defense claimed there was no gathering of anti-unionists, no subsequent pursuit, no guns drawn, and no shooting. Even Gastonia police officer William Whitlow, who witnessed the truck turn around in Gastonia and head back to Bessemer City, testified that there was no mob, adding, "There was nothing to excite my curiosity." He also denied that the driver of the truck was threatened bodily harm by defendants Jones and Lunsford if he did not turn around, as the prosecution claimed.[31]

The other two defendants, Lunsford and Wheelus, who were identified as being in other vehicles in the pursuit, claimed not to be at the scene at all. The accident was reported to have happened at 2:45 p.m. and Lunsford claimed to be at home sleeping until 3:00 p.m. He said that soon after rousing he heard there was a wreck and he investigated. He testified that he found defendant Morrow and truck driver Lingerfelt arguing over who caused the accident. He said he witnessed no shooting. Wheelus claimed to be in Charlotte at the time of the shooting and his wife supported this claim on the stand. A number of individuals supported Wheelus's alibi, including Loray Mill superintendent O.G. Morehead, Loray Mill resident agent J.A. Baugh, Jr., and even a minister, the Reverend A.E. Hill. These witnesses and others claimed that Wheelus was present at the court investigation of the September 9 kidnapping and flogging of three NTWU members that was held in Charlotte the day of the shooting. One of those accused of the kidnapping and flogging, Dewey Carver, claimed that Wheelus drove him to the investigation himself. Carver testified that Wheelus drove him back after the court adjourned at 3:15 p.m. Wheelus said that upon arriving in Gastonia he heard about the wreck and arrived at the location at 4:00 p.m., when Solicitor Carpenter was investigating the incident.[32]

Striker D.L. Case testified to what he witnessed the day Ella was shot. He said that he was on the strikers' truck when it all occurred. He said Horace Wheelus was responsible for Ella's death and he was certain Wheelus was the one, because like Wheelus the shooter had three fingers missing from his left hand. In the courtroom during testimony, Wheelus tried to cover his left hand with his hat. Case also identified the gun as "a blue steel, medium sized pistol." Julius "Crip" Fowler, a one-legged textile worker who was also on the truck, said he too saw Wheelus point and fire directly at Ella May. Willie Hilburn, another passenger on the truck, said he saw Wheelus fire and Ella fall immediately after. Numerous others, some union men and women and others unaffiliated with either side who happened to be near the highway, placed Wheelus at the scene. One noted that, while passing onlookers during the car pursuit, Wheelus pulled his hat over his head in such a manner as to avoid identification.[33]

W.O. Bradley was standing next to Ella in the bed of the truck and testified that Wheelus was a stumbling drunk at the time of the shooting. Mrs. Noah Ledford was Wheelus's neighbor and said that he had asked her to give him an alibi during the investigation. She refused and replied, "I won't swear to a lie for no man and if I go to court, I'll swear the truth that you left home between 10 and 11 o'clock in the morning and didn't come home till 7:30 or 8." She also said that when he returned home that night he was so drunk that he parked his car in her garage by mistake. Only one member of the Committee of 100 openly spoke at the trial about its activities, and that was H.W. Branch. His evidence was later ruled inadmissible.[34]

Bulwinkle made closing arguments for the defense and insisted that the jury "remember that these defendants come into court with clean hands, men who uphold the majesty of the Constitution, while this other crowd goes throughout the land seeking to overthrow the government—the very riff-raff of civilization." On March 6, 1930, all the defendants were acquitted after 35 minutes of deliberation. The *Gastonia Gazette* printed in its March 7 paper that the jury "turned to the more reasonable explanation." This "more reasonable explanation" is that Ella May was shot by a fellow striker or the driver of the truck. This "more reasonable explanation" relied on conflicting testimony between defense witnesses. Some claimed there was no violence whatsoever. Others claimed there was violence, but they were unable to identify the guilty parties.[35]

The 100 witnesses who took the stand for the prosecution outnumbered the defense witnesses 2:1. Many of these witnesses successfully identified specific individuals who were armed and shooting at the scene. From testimony given it is apparent that there was a mob blocking the strikers, there was a car chase, and there was a mass shooting. It is also reasonable to believe that Wheelus was the man who shot Ella May. Despite the evidence, no one was convicted, and no one was ever held criminally accountable.[36]

The majority of the information collected here about the murder trial was obtained in the *Gastonia Gazette*. The newspaper printed day-

NTWU members convicted of Aderholt's murder. Front, left to right: William McGinnis, "Red" Hendricks, Joe Harrison. Back, left to right: Louis McLaughlin, George Carter, Fred Beal, Clarence Miller (courtesy Millican Pictorial History Museum).

by-day recaps of the trial as events unfurled. Additional revelations would be found in the case transcripts, but unfortunately they have mysteriously disappeared from the court record. No doubt officials were eager to wipe away the details of such a case, erasing the evidence from future generations.

Ella's brother Wesley hired a detective to investigate his sister's death but ran out of funds before the investigation was complete. It seemed as if justice would never be served. A last effort was made by family to hold those responsible accountable. On December 12, 1929, civil lawsuits were brought against 21 individuals involved in the shooting as well as the Manville-Jenckes Company, the Meyers Mills, and the *Gastonia Gazette* for a total of $200,000. The charges included an unlawful and violent conspiracy to prevent Ella May from attending a legal union meeting. The civil lawsuits were brought on behalf of Ella's brother Wesley, who was the administrator of the estate of Ella May for the benefit of her children. The American Civil Liberties Union assisted Wesley in filing the claims. In June of 1930 the suits for damages in Ella May's case were heard and thrown out of court by the judge before they were brought to trial. No one was ever held responsible for Ella May's murder.[37]

Textile workers' lives did not improve. In October 1929, the stock market crashed and demand for cotton goods further diminished. Continuation of new methods such as the stretch-outs and owners' investments in labor-saving machinery softened economic losses somewhat. However many mills across the country were shut down, declaring bankruptcy, and most that remained running did so at reduced hours. Profits for owners were almost nonexistent from 1930 to 1933. By 1930 only about half of Gastonia's townspeople were employed. The Manville-Jenckes Company had to make drastic cuts and declare bankruptcy. At that point about 400 of its around 600 mill homes in Gastonia were empty. Even if momentum had existed for strikers following Ella's death, it would have been wiped out by the backlash of the nation's economic situation, the Great Depression.[38]

In 1930 there was no open NTWU activity in Gaston County and most workers were working for even less pay. The NTWU Gaston County strikers were in no better condition than before the union organized. The only hint that the union had even existed was its building headquarters, and even there the Committee of 100 had declared victory by displaying an American flag on it. The *Gastonia Gazette*, which was so vocal during the strike, made no mention of it during its anniversary edition in 1930. While at the time the significance of the strike was clearly felt locally, nationally, and even internationally as the events unfolded, the community in which the historic episode occurred was quick to forget it had even existed.[39]

The mill owners never acknowledged any responsibility or regret regarding Ella May's death. The *Southern Textile Bulletin*, a mill industry journal, published the following about her: "Ella May Wiggins was not a widow but a woman whose husband had left her. Her reputation was so bad that she had, long before she ever heard of Communism, been requested to move out of several mill villages. She was a 'hard boiled' type of woman who loved a quarrel and a fight, and her home had been the scene of many disorders.... There is no use trying to paint Ella May Wiggins as a saint or a woman whose character was beyond complaint."[40]

On April 11, 1930, the *Gastonia Gazette* also tarnished Ella May's character. "The poor woman is dead and gone, and it is not for anyone to say much about her now, but if some of these sob sisters who are writing about her 'riding into the West' and all that sort of ridiculous tommyrot, knew the real life story of Ella May, there would be a different story. Her children, motherless and nameless, are far better off where they are than if their mother were living."[41] Ella's children would have disagreed strongly with those statements, especially the last. On the occasions they did discuss the painful loss of their mother, their regret at losing her and the resentment they had towards the loveless childhood they received in the orphanage and relatives' homes were always in the forefront.

Discontentment in Ella's mill and the nearby American Mill No. 1 grew exponentially after her death. In August 1930 a spontaneous textile strike broke out at both the American Mill No. 1 and the American Mill No. 2 in Bessemer City. From December 1929 to August 1930 there had been three separate salary cuts, the last of which cut textile workers' pay by thirty percent. Workers were not able to cover their most basic needs. Men and women working at the American Mills were desperate. NTWU and AFL representatives rushed to organize the strikers. Both the NTWU and AFL affiliates met hostile workers who ran them out of town. The men and women who had so recently been abandoned by the NTWU and experienced the murder of their native songstress wanted nothing to do with any union. Community members were distraught over the dispute between the workers and the owners in the American Mills. They feared violence. The community was more compassionate towards the strikers' plight in this situation than in previous strikes. The community blamed the owners, the Goldbergs—who were already paying the lowest wages in Gaston County, then cutting back further—as the cause of the revolt. The Reverend C.J. Black, the Baptist minister who presided over Ella May's funeral, led a mediation committee of local leaders. The committee mediated between the Goldbergs and the unrepresented strikers. After a week the Goldbergs agreed to roll back the most recent pay cuts but refused the request to reduce rent in the mill villages. Strikers agreed to the compromise and the strike ended.[42]

Communism, like striking activity, was not completely eradicated from the South. On January 1, 1930, the region was divided into two districts by the Communist Party. One headquarters was in Birmingham. The second headquarters fluctuated between Charlotte, Greensboro, and Winston-Salem. The day in March that those charged with Ella's murder were not indicted, 500 to 800 whites and blacks gathered in Charlotte after the ruling to hear ILD representatives speak. Representatives spoke against high unemployment and urged the mixed crowd to unite under the Communist Party. Propaganda literature continued to be distributed. A black tenant farmer accused of attacking a white woman was lynched in 1930 in Tarboro, North Carolina. Charlotte Communists distributed flyers alleging that a capitalist mob of landlords and storekeepers had murdered the man. They compared this gang violence to the attack on Ella May. The Communists called for black and white workers to unite against the bosses' tyranny. A flyer stated, "It is clear now that terrorism and lynching is being organized by the mill owners and their agents." The communist organization continued to protest lynching in the South throughout 1930. The death penalty for lynching was proposed. Twenty-five whites and fifteen blacks met to rally against lynching in Charlotte. The focus on lynching drew black members to the Communist Party's American Negro Labor Congress (ANLC). ANLC chapters grew from five Northern chapters to thirty-four Northern and twelve Southern chapters. Despite Communist leaders' efforts to integrate black and white workers, they intermingled rarely and in small numbers.[43]

However, counterefforts to unite white workers against the competition of black labor spread and gained momentum during the Depression. When work was plentiful, white workers prided themselves on having higher class jobs. However, during the Depression, driving trucks, sweeping floors, digging ditches, and serving as a bellboy were no longer looked down upon as "nigger work." Any job was an opportunity and many white men didn't want black men keeping those opportunities away from them. Black-shirt rallies were held in 1930 where black shirts were sold and forms were signed that read, "If you are opposed to social equality and amalgamated marriages; and if you are in favor of white supremacy and the enforcement of law, against Communism in the United States, then you are by nature and principle a Black Shirt." Black Shirt members denied black workers' rights to work anywhere but in agriculture. The Black Shirt organization exploded in Atlanta, where within weeks twenty-seven thousand members joined.

Black Shirts boasted, "Before Christmas there isn't going to be a black bellboy or black truck driver in Atlanta." The Black Shirt organization's appeal was so far-reaching that they ironically even enticed some NTWU members to a meeting in Charlotte.[44]

No doubt many former Committee of 100 members who were united in ousting the NTWU united behind the Black Shirt organization. It isn't surprising that the Black Shirts rallied behind the alleged leader of the Committee of 100 when he ran for Congress. Major Bulwinkle, Loray Mill lawyer and alleged leader of the Committee of 100, ran for Congress in 1930. He announced his bid for office the day after he successfully helped clear those accused of Ella May's murder. The *Gastonia Gazette* printed in its March 8 paper that Bulwinkle had a "fine record" and that "for the past year Major Bulwinkle has devoted his time and talent largely to fighting the Communist movement in this section. In so doing he has rendered a great service to this part of the country and his record in this respect will no doubt do much to forward his candidacy." Not only the Black Shirts but a mix of organizations with anti–Communist leanings, including the state AFL and the Charlotte Central Labor Union endorsed him. Bulwinkle ran against two former NTWU members, lawyer Tom Jimison, and Gastonia textile worker and striker Dewey Martin. Bulwinkle won in a landslide and was sent to Washington.[45]

The Communist Party continued to fight injustice against blacks. The ILD came to the defense of the Scottsboro Boys in 1931. Nine black youth ranging in age from thirteen to twenty were pulled from a train in Alabama after a brawl between white and black men was reported. Two white women dressed in men's overalls who had jumped the train were apprehended with the white and black men who were discovered. Fearing such charges as vagrancy, drunkenness, and prostitution, the women claimed to be victims. They alleged the black men raped them. One of the women later recanted her story. Despite the young woman's retraction and the ILD's support, the young men were found guilty.[46]

The National Industrial Recovery Act (NIRA), which was designed to ignite a national economic recovery, was signed by President Franklin D. Roosevelt in 1933 as part of his "New Deal." The act encouraged manufacturers to increase employment, stimulate consumer consumption, shorten working hours to forty hours a week, raise wages to twelve dollars a week, and eliminate child labor. NIRA also acknowledged workers' rights to organize. These government guidelines gave hope to workers that their situation would radically and rapidly improve. Anticipation of improvement was soon replaced by disappointment. Owners and management found methods of chipping away at the act. Complaints made to the National Labor Board (executive body of the NIRA) by textile workers were referred to the Cotton Textile National Industrial Relations Board (their trade association). From August 1933 to August 1934, 3,920 complaints were made, 96 investigations were held, and only one investigation manifested into favorable improvement for textile workers' wages and hours. Roosevelt's attempt to solve what he called the country's "number one economic problem," Southern poverty, floundered.[47]

In 1934 a final and all-encompassing wave of strikers referred to as the General Strike hit the textile industry. The walkouts initiated in Alabama and spread like wildfire throughout the South. The strike extended as far as New England. Flying squadrons, hundreds of strikers in cars and trucks, traversed the countryside and moved from town to town. They shut down unprepared textile mills at an unprecedented rate. Three hundred thousand employees fought for various improvements such as a $12 minimum wage for a forty-hour work week, elimination of the stretch out, and union recognition. This time the United Textile Workers Union organized the majority of the strikers and virtually every mill in the Carolinas was shut down. For twenty-two days, hundreds of thousands of workers and countless weaving looms stood idle. Tension filled the streets again in Gaston County and a real fear of violence manifested. Twenty-two mill executives signed a petition pleading for government intervention that ended: "[U]nless something is done within the next twenty-four hours no authority can be held responsible for results." The state militia was called and strikers were pushed back. In Georgia the governor interned flying squadron participants in a hastily put together detention camp. Southern governors, textile mill owners, and the National Guard together worked at ending the largest strike, in both numbers and breadth, in American history.[48]

Feeling broken and disenfranchised, workers returned to work again, gaining nothing but additional heartache. Never again in the South would the word "union" bring a wide-reaching reaction of uniting hope to textile employees. Better off to suffer in silence than to stir up trouble that would lead to additional suffering and inevitably end in defeat. The word "union" now has a sour taste in the Southern labor climate, not just in the textile industry, but in industry in general. The anti-union sentiment runs deep, as an almost universal cultural value whose cause and origins are long forgotten, but the value itself remains entrenched. To this day unions of any type are a rarity in the South, especially in the Piedmont. The NTWU, which was created to unite Southern textile mills and the Southern industry in general, was dismantled altogether in 1934. Leaders of the Communist Party in Russia directed American members to join their old nemesis the AFL.[49]

Firestone, which had been a major purchaser of tire fabrics from Manville-Jenckes, bought the Loray Mill in 1935 and continued producing tire cord there until 1993. In 1987 labor conflict erupted again in the old Loray Mill over production quotas and recently cut pay. The United Rubber Workers Union that represented Firestone employees elsewhere made an effort to organize. The memories of the 1929 violence were reawakened at this time and highlighted in newspaper articles. A vote was held by employees and the idea of unionization was rejected. The fear of another brutal conflict was blamed for the snub.[50]

In 1938 the Fair Labor Standards Act was established to extinguish labor conditions harmful to the health and efficiency of workers and unfair methods of competition based on these conditions. A minimum wage of twenty-five cents was enacted and the payment of time-and-a-half was required for hours over forty-four in a given workweek. It also forbade any labor involving children under sixteen except for those children between the ages of fourteen and sixteen who still attended school. The Fair Labor Standards Act improved pay and work conditions for textile workers and further cut away at mill owners' ability and willingness to invest in their workers' personal lives. Company housing was seen as a drain of resources. In the years following the act, owners sold more and more of their nonessential company property. Work and social relations were separated. Often workers welcomed increased pay and better working conditions, and enjoyed having their lives controlled less by mill bosses.[51]

During World War II, the textile industry experienced a temporary boom. The mid-forties were a time of prosperity while supplying the armed forces with cotton goods. Uniforms, tents, and other essential

176

items were needed in vast quantities. Over 12 billion yards of cotton were ordered each year during the war. Ella May's grandson Bobby Bolch, Myrtle's son, benefited from the increased demand for cotton goods. He lived in Gastonia and worked at the Loray Mill as a teenager. He worked in several mills throughout his lifetime. Throughout the years a number of Ella's relatives relied on the textile industry for subsistence, her brother Wesley for one. To the textile mills' credit, especially the Loray Mill, no discrimination against Ella May's relatives kept them from employment. In fact, Ella May's grandson Victor Wiggins, Albert's son, in the early '70s, while waiting in an employment line outside of Loray, was asked by mill management if he was related to Ella May, and after responding yes he was given a job that very day.[52]

The 1970s, while a progressive time for the industry, was a tough decade for its workers. Modernization of equipment reduced the numbers of workers needed. International competition strained management to find a way to make profits. "Made in the USA" became a rallying call to the public to support its American workforce. Inevitably the conditions that enabled the rise of the textile mills in the South were the same that caused the industry's demise. Cheap labor in developing countries, along with low costs for production, moved investors' interests abroad. Communities whose birth and development centered on textiles had to diversify their economies. More and more textile mills became vacant. The immense brick buildings, once centerpieces of their communities, slowly became deteriorating and forgotten ghosts of their former selves. The Piedmont's textile heritage was forgotten by many. Both the pride in the growth of the "City of Spindles," as well as its pain and struggle to survive, faded into history.[53]

Labor organizations made efforts to keep the memories of past labor struggles alive. In 1977, Charlotte union leaders paid tribute to Ella May at a graveside memorial service. The Charlotte Central Labor Council President William Brawley spoke of plans to see a monument built in her honor. The statement "Heroine of the bloody Loray Mill strike of 1929" was to be inscribed on her marker. The monument was to replace the simple stone that Wesley placed at the site nearly fifty years prior.[54]

Daughter Millie was fifty-five and had not been notified of the events that took place. She was not upset at the fact she was not invited to partake in the union gathering. She felt bitter towards unions and blamed the NTWU for her mother's death. Upon hearing the labor president's plans she said, "Are they trying to do something for my mother or are they trying to feather their nest?"[55]

At home in New York, Millie was interviewed about the tribute held in Bessemer City by a local paper. The article was titled "Tribute Rekindles a Nightmare." Millie, surrounded by family, retold the tragedy of her mother's death. Pain and pride were evident when she talked about her mother. Resentment and suspicion were evident when she talked about unions.[56]

Suspicion of motives behind those remembering and honoring Ella May were felt not only by Millie but by a number of Ella's descendants. Ella May's murder is often portrayed as a martyrdom for a cause. To children who had a mother taken away, and to grandchildren who lived with the absence of a grandmother, "the cause" is resented to varying degrees, for it took a loved one away. Daughter Charlotte explained, "You've got five children that grew up to be very bitter. We didn't know what a mother's love was." Those descendants more removed, and less directly touched by the loss of Ella May, such as her great-grandchildren, have embraced public recognition of Ella May's contribution to history more readily.[57]

On September 15, 1979, fifty years and a day after Ella May's murder, a tribute was held to honor her. The National Organization of Women (NOW) and the Charlotte and Hickory Central Labor Councils sponsored the event and it was called "Ella May Wiggins Day: A Tribute to Women Textile Workers." A granite cross with the words "ELLA MAE WIGGINS: She was killed carrying the torch of social justice, September 14, 1929" was placed on her still unmarked grave. A variety of contributors coordinated by the Charlotte Central Labor Council sponsored the marker. Frank Sisk, the funeral home director who had protected Ella's body and children years earlier, was present. So too was Lynn Barnett, a fellow Bessemer City striker who was in the bed of the truck with Ella when she was shot.[58]

Three of Ella May's children, Millie, Charlotte, and Albert, participated in the remembrance. Millie was hesitant about participating. She was the only sibling who agreed to go who was old enough to remember the painful events that led to their mother's demise. During the remembrance, each of Ella's children was individually called to be recognized. When Millie's name was called she remained seated, uncomfortable with the idea of being put on display. Only after her name was called repeatedly and she received words of encouragement from her daughter Carmen did she finally stand up and allow herself to be publicly recognized by the crowd as a daughter of Ella May.[59]

Ella's sister-in-law Mary May, Wesley's wife, was also one of the two hundred people gathered. Ella's brother Wesley, a then seventy-two-year-old retired Gastonia textile worker, chose not to attend. Even fifty years later the memories were too painful. He also felt resentment towards the new marker and protective of the simple stone he had placed at his sister's gravesite fifty years prior. There had been talk of removing it. Upon news of this, Wesley told family members, "I hope it breaks their damn back[s] pulling that stone up." After some discussion by event organizers, the original rock remained. A slide show titled "Textile Women Fight Back" was shown at the event. Later, textile songs were sung, including the militant union anthem "Solidarity Forever." None of Ella May's ballads were sung that day. The event began in the American Legion Hall of Bessemer City, which was the same hall that had deputized men the day of Ella's murder. A booklet of songs and information on Ella May was distributed to participants. Inside it, Vera Buch wrote a memorial to Ella May:

> Dear Ella Mae:
>
> It is fifty years now since you and I were together here in Bessemer City, fighting for the National Textile Workers Union. You had a life of grinding poverty with nine children to support. The strike gave you the opportunity to blossom out, to be a human being, to show your great courage and intelligence. You saw from the start the importance of our union standing for black and white uniting in the struggle against the mill owners and their agents, the state forces. You went from door to door among the black people getting them into the union. You organized that meeting in Stumptown where I and Albert Weisbord, leader of the union, spoke. It was for this they killed you, Ella Mae.
>
> Vera Buch Weisbord
> Chicago, Illinois
> September 14, 1979[60]

Ella and other Gaston County textile workers were paid homage to again in June of 2004. "History Happened Here: The Many Stories of

Ella May's Granite headstone and simple stone placed by Ella's brother Wesley (photograph by the author 2007).

Loray" was the title of the symposium held on the seventy-fifth anniversary of the Loray Mill Strike. The community-initiated event was held at the Gastonia Adult Recreation Center, formerly the Gastonia Armory. Gastonia local Lucy Penegar, a longstanding proponent of saving the Loray Mill, spearheaded the well-received one-day summer event. It was marked by community members and historians sharing the community's textile heritage, the Loray Mill Strike's history, and open group discussions of various perspectives of the strike. One breakaway session was titled "Stories of Ella May and Women Workers." A handful of Ella's descendants participated in this and other open group discussions. Mill music, including Ella May's famous "Mill Mother's Lament," was performed. Traditional mill lunches were also available for purchase. A scene from *Finding Clara*, a play based upon the Loray Mill Strike, was shared. I portrayed my great-grandmother in the scene. The day-long affair with about four hundred participants ended with a tour of the old Loray Mill.[61]

The infamous Loray Mill, for years known as the Old Firestone Plant, is still standing. For decades it sat aging and vacant. Preservation North Carolina received possession of the property in 1998 after Firestone donated the property to the city. In 2001 the mill and surrounding village housing became part of the National Register of Historic places. Despite the old mill's deteriorated state, it was still majestic in stature and had an eerie aura about it that hinted at its painful past. For years the decision between tearing down the worn building or developing the property was hotly contested in the community. Many saw the mill as an eyesore with little hope of improving the community's image historically or fiscally. Others saw the preservation of the historic landmark as an opportunity to serve the community while paying homage to the past. After years of debate and delay, in the summer of 2012 the paperwork was signed and the process of restoring the old mill to its former glory began. In 2015 the Loray Mill opened its doors again to the community, donning its former name. Like other restored mills, the Loray Mill now has an industrial feel and unique archaeological features that intermingle with modern interior design and purposes. The property now offers affordable loft housing and commercial space. The Loray Mill's history is not only found in the design of the building itself, but the carefully selected textile artifacts and artwork that dot the mill's common areas.[62]

Top: Old Firestone Mill (Loray Mill) (photograph by the author 2007). *Bottom:* The restored Loray Mill, 2015 (photograph by Tammy Cantrell).

Ella May's Bessemer City mill, American Mill No. 2, like many mills was foreclosed on during the Great Depression. In 1935 it closed and then in 1936 it was reorganized by the same ownership, the Goldbergs, who renamed the mill Algodon No. 2. Only a year later the mill was sold by the Goldbergs and the machinery was liquidated. Pyramid Mills,

Modern-day mill house in front of unrestored Loray Mill in 2011 (photograph by Leo Hohmann).

2013 American Mill No. 2, now Dawn Processing (photograph by Tammy Cantrell).

owned by M. Rauch, eventually took over the idle plant and kept the name Algodon and ran it for about 30 years. The mill was then purchased by Henry Moore in 1963 and became known as Dawn Processing Company.[63]

The mill today still stands, old and deteriorating, one of countless others in Gaston County that survived being torn down, yet not one of the few lucky enough to be restored. It is a smaller mill with no particularly striking features. However, it was the workplace of a woman who took a stand. As a result of her stand she is one of a select few women in the United States who were assassinated for their beliefs. So the little mill that seems to be forgotten has a special place in our nation's history.

The North Carolina state advisory proposed a marker in 1986 commemorating the Loray Mill Strike. The sign was to be placed near the mill on Gastonia's main street, Franklin Boulevard, and say, "Loray Strike—A strike in 1929 at Loray Mill one block south left two dead and spurred opposition to labor unions statewide." After some debate the marker was rejected. The idea of being reminded of the violent confrontation that divided the community and resulted in two murders did not sit well with many. Gastonia City Council members proposed a revised version that identified the community's defeat of the first communist threat to control Southern textile mills. This revision was rejected by the state committee. The marker proposal was reopened in 2007 after Gastonia officials and Preservation North Carolina asked the state to do so. This time there were no objections and the original inscription was made. Preservation North Carolina waited until the restoration work on the Loray Mill began before placement. As a result the marker waited in storage for 5½ years. On April 28, 2013, the marker acknowledging the Loray Mill Strike was finally displayed in Gastonia for the very first time. On this rainy Sunday, I was standing with the crowd to witness the unveiling and was surprisingly given the honor of helping unveil it.[64]

A marker commemorating Police Chief Orville Aderholt stands near a bridge in Gastonia giving homage to his service and early demise. No such monument yet stands for Ella May. Ella May is currently commemorated at other locations in the community. Gaston County's Museum of Art and History has some information regarding Ella May in their "Carolinas Textile Exhibit: The Ties that Bind." Charlotte's Levine Museum of the New South has included a short excerpt about Ella May in its permanent installment "From Cotton Fields to Skyscrapers." The Textile Heritage Museum at the Glencoe Mill Village near Burlington, North Carolina, includes Ella May in "Neighbors Divided," part of a large installment called "Rhythm of the Factory Series of Markers."[65]

As of April 28, 2013, the Loray Mill Strike has a historical marker (photograph by Tammy Cantrell).

Ella May never was forgotten. She was credited with inspiring the late singer and songwriter Woody Guthrie. Pete Seeger too was touched by Ella May's story and performed in my local community when I

185

Orville Aderholt's great-nephew Dennis Aderholt (on the right) poses with Gastonia police officers in 2004 during the Aderholt Memorial Bridge dedication. Dennis is a captain with the Mecklenburg County Sheriff's Office Reserve Unit (courtesy Gastonia Police Department).

was nine years old. He personally invited my sixty-two-year-old grandmother Millie to the concert and she proudly attended. He played several of Ella's original ballads. "Mill Mother's Lament" was recorded on Pete Seeger's 1956 *American Industrial Ballads* album that was later preserved in Smithsonian Folkways. John Greenway also recorded Ella May's "Chief Aderholt" on his *American Industrial Folksongs* album in 1955. A number of pamphlets, articles, and books have published Ella's songs, most prominently her most popular ballad, "Mill Mother's Lament."[66]

At least six fictional novels based on the events that took place at the Loray Mill Strike were written in the 1930s: Mary Heaton Vorse's *Strike!* (1930); Sherwood Anderson's *Beyond Desire* (1932); Fielding Burke's (Olive Tiffard Dargan) *Call Home the Heart* (1932); Grace Lumpkin's *To Make My Bread* (1932); Myra Page's *Gathering Storm* (1932); and William Rollins's *The Shadow Before* (1934). Each had a fictional character based upon Ella May. Vorse and Lumpkin had personal insight developing their Ella May characters, having met Ella in person during the Loray Mill Strike. As well as writing *To Make My Bread*, Lumpkin published articles in the *Nation* and the *New Masses* about some of the songs Ella wrote. Vorse and another author, Tom Tippett, who wrote *When Southern Labor Stirs*, were journalists who covered the strike, as well as the following trials, and later wrote about them.[67]

Today there are plans for additional plays, an opera, and another book being written. The Ella May Wiggins Memorial Committee now has a website, ellamaywiggins.org, and its Facebook page with nearing 1,000 members is active and growing. Currently the nonprofit organization is working towards raising money for a life-sized bronze statue of Ella May. Ella May's memory is very much alive and her recognition will continue to grow.

I hope that this book not only helps preserve Ella May's memory, but encourages others. Our ancestors are our roots, the source from which we develop. In order to fully understand ourselves we must understand these ancestors who stood before us. Our roots are often buried. It is our responsibility to dig, to ask questions, to discover our families' histories. We all need to seek out our families' stories and then become the storytellers for future generations. When we accomplish this, not only do we better understand where we came from, but we better understand who we are, and help our own descendants to do the same. Beyond an understanding of who we are is an understanding of who we can be. Our history can help us unearth this as well. For example, Ella was independent, resourceful, aware, just, courageous, and tenacious. These are qualities to aspire to have. We honor our ancestors not only by remembering them, but by sharing their stories, and by aspiring to be the best of who they were.

Appendix

The Ballads

LITTLE MARY PHAGAN (1913)

This particular version originates from the Tennessee Mountains, near the North Carolina line, from Bertha Bailey. It is found in *North Carolina Folklore*, Volume II. © 1952, Duke University Press. All rights reserved. Republished by permission of the copyright holder. www.duke upress.edu.

Little Mary Phagan she went to town one day.
She went to the pencil factory to see the great parade.
She went to draw her money that she had worked the week before,
Down in the pencil factory with Leo Franks, you know.

She left her home at eleven o'clock and kissed her mother good-bye;
Not one time did the poor girl think that she would have to die.
Leo Franks met her with a bruteless heart, you know.
He smiled and said, "Little Mary, you will go home no more."

He sneaked along behind her till she reached the middle room.
He laughed and said "Little Mary, you met your fatal doom."
Down on her knees she was crying, "Leo Franks," she pled.
He took a stick from the trash pile and struck her across the head.

While the tears rolled down her rosy cheeks and the blood flowed
    down her back,
She remembered telling her mother the time she would be back.
How Leo Franks killed little Mary, it was on one holiday.
He called on old Jim Conard for to carry her body away.

He drug her down the stairway by her head, by her feet.
Deep down in the basement little Mary lay asleep.
Utely was the watchman. He came to wind his keys.
Deep down in the basement little Mary he could see.

He called upon the officers, some names I do not know.
They come and said, "Jim Conard, Jim Conard, you must go."

189

They took him to the jailhouse and locked him in the cell,
But the poor old innocent negro knew nothing for to tell.

Little Mary is in heaven, and Leo Franks in jail,
Waiting for the day to come that he could tell his tale.
But the poor old judge and jury they passed his sentence well.
Little Mary is in heaven, and Leo Franks in hell.

SOLIDARITY FOREVER—*Ralph Chaplin (1915)*

This song is sung to the tune of "Battle Hymn of the Republic."
This version is found in *Songs of Work and Protest*.

When the union's inspiration through the workers' blood shall run,
There can be no power greater anywhere beneath the sun;
Yet what force on earth is weaker than the feeble strength of one,
But the union makes us strong.

CHORUS: Solidarity forever,
    Solidarity forever,
    Solidarity forever,
    For the union makes us strong.

Is there aught we hold in common with the greedy parasite,
Who would lash us into serfdom and would crush us with his might?
Is there anything left to us but to organize and fight?
For the union makes us strong.

It is we who plowed the prairies: built the cities where they trade;
Dug the mines and built the workshops, endless miles of railroad laid;
Now we stand outcast and starving midst the wonders we have made;
But the union makes us strong.

All the world that's owned by idle drones is ours and ours alone.
We have laid the wide foundations; built it skyward stone by stone.
It is ours, not to slave in, but to master and to own.
While the union makes us strong.

They have taken untold millions that they never toiled to earn,
But without our brain and muscle not a single wheel can turn.
We can break their haughty power, gain our freedom when we learn
That the union makes us strong.

In our hands is placed a power greater than their hoarded gold,
Greater than the might of armies, magnified a thousand-fold.
We can bring to birth a new world from the ashes of the old
For the union makes us strong.

A SOUTHERN COTTON MILL RHYME

The origination of this ballad is unknown. This version was submitted to the *New Masses* by Grace Lumpkin in 1930. It is found in *Wobblies, Pile Butts, and other Heroes.*

I lived in a town away down south,
By the name of Buffalo,
And worked in a mill with the rest of the trash,
As we were often called you know.

You factory folks who read this rhyme,
Will surely understand,
The reason why I love you so,
Is I'm a factory hand.

While standing here between my looms,
You know I lose no time,
To keep my shuttles in a whiz,
And write this little rhyme.

We rise up early in the morn,
And work all day real hard,
To buy our little meal and bread,
And sugar tea and lard.

We work from week's end to week's end,
And never lose a day,
And when that awful pay day comes,
We draw our little pay.

We then go home on payday night,
And sit down in a chair,
The merchant raps upon the door,
He's come to get his share.

When all our little debts are paid,
And nothing left behind,
We turn our pockets wrong side out,
But not a cent can find.

We rise up early in the morn,
And toil from sun till late,
We have no time to primp,
And fix and dress right up to date.

Our children they grow up unlearned,
No time to go to school,
Almost before they've learned to walk,
They learn to spin or spool.

The boss men jerk them round and
Round and whistle very keen,
I'll tell you what, the factory kids,
Are really treated mean.

The folks in town who dress so fine,
And spends their money free,
Will hardly look at a factory hand,
Who dresses like you and me.

As we go walking down the street,
All wrapped in lint and strings,
They call us fools and factory trash,
And other low down names.

Just let them wear their watches fine,
And golden chains and rings,
But when the day of Judgment comes,
They'll have to shed those things.

UNION BOYS—*Kermit Harden (1929)*

The song is sung to "Johnny Boy." It is found in *Strike Songs of the Depression*.

Crowd around me here, union boys,
And lend me your ears, union boys.
You've a way of knowing
I've a way of showing
What the union means, union boys.

Beal was sent from Bedford and we know his worth
He will make a heaven for us here on earth.
We've all worked for low pay right here in Loray.
Our union will stay, union boys.

Manville Jencks betrays us, but the workers all stand by us,
We still have the union, union boys.
And when we get our high pay, we will never, never stray
From our eight-hour day, union boys.

T.A. Smith will squeal and squeal
Painter he will steel and steel.
With our union we will go, union boys.

The Gazette is against us,
Pershing is with us. Choose of the two, union boys.

Smith will grow lonely, want us and us only.
With our union we will go, union boys.

So climb up on the ties, union boys.
Show up all their lies, union boys.
Jencks may lie and cheat us, but he'll never beat us
Until we've won our strike, union boys.

WE NEED YOU MOST OF ALL—*Russell Knight (1929)*

This is set to the tune of "I Love You Most of All." It was originally printed in the *Daily Worker* and then later reprinted in *Strike Songs of the Great Depression.*

The time has come for our freedom
We must stand up and fight
The strike is on, boys, stick to it
And we will win out all right.

Let's keep the darn mill standing
No matter what they all say
Then we will soon see a-dawning
For us, a brand new day.

We must look to our union then
Stick with it through thick and thin
Then you will hear the boss let out a cry—
"If you don't give in I'm going to die."

So just be patient with Pershing and Beal
Until old Baugh lets out his squeal,
"Please come back, we'll recognize you
As a union through and through."

So to the union we send our call
For we need you most of all.

CHIEF ADERHOLT—*Ella May (1929)*

The music for "Chief Aderholt" is from the 1925 ballad "Floyd Collins." This ballad was printed in the National Organization of Women's 1979 pamphlet *Let's Stand Together.*

Come all of you good people, and listen while I tell,
The story of Chief Aderholt, the man you all knew well.

It was on one Friday evening, the seventh day of June,
He went down to the union ground and met his fatal doom.

They locked up our leaders, they put them into jail,
They shoved them into prison, refused to give them bail.
The workers joined together, and this was their reply,
"We'll never, no, we'll never, let our leaders die."

They moved the trial to Charlotte, got lawyers from every town,
I'm sure we'll hear them speak again upon the union ground.
While Vera, she's in prison, Manville-Jenckes is in pain,
Come join the Textile Union, and show them you are game.

We're going to have a union all over the South,
Where we can wear good clothes, and live in a better house.
Now we must stand together, and to the boss reply,
"We'll never, no, we'll never, let our leaders die."

Ild Song—*Ella May (1929)*

The music for this particular call of support from Ella is taken from a common hymn, "Weary Pilgrim on Life's Pathway." This song was also published in *Let's Stand Together*.

Toiling on life's pilgrim pathway,
Wheresoever you may be.
It will help you, fellow workers,
If you will join the ILD.

CHORUS: Come and join the ILD
Come and join the ILD
It will help to win the victory.
If you will join the ILD.

When the bosses cut your wages,
And you toil and labor free.
Come and join the textile union,
Also join the ILD.

Now our leaders are in prison,
But I hope they'll soon be free.
Come and join the textile union,
Also join the ILD.

Now the South is hedged in darkness,
Though they begin to see.
Come and join the textile union,
Also join the ILD.

ALL AROUND THE JAILHOUSE—*Ella May (1929)*

Ella modeled this song on "All Around the Watertank, Waiting for a Train." It is found in *Let's Stand Together*.

All around the jailhouse
Waiting for a trial;
One mile from the union hall
Sleeping in the jail.

I walked up to the police man
To show him I had no fear;
He said, "If you've got money
"I'll see that you don't stay here."

"I haven't got a nickel,
"Not a penny can I show."
"Lock her up in the cell," he said,
As he slammed the jailhouse door.

He let me out in July,
The month I dearly love;
The wide open spaces all around me,
The moon and stars above.

Everybody seems to want me,
Everybody but the scabs.
I'm on my way from the jailhouse,
I'm going back to the union hall.

Though my tent now is empty
My heart is full of joy;
I'm a mile away from the union hall,
Just a-waiting for a strike.

TWO LITTLE STRIKERS—*Ella May (1929)*

Ella May's ballad "Two Little Strikers" follows the melody of "Two Little Children." It is found in *Let's Stand Together*.

Two little strikers, a boy and a girl,
Sit by the union hall door.
The little girl's hand was brown as the curls
That played on the dress that she wore.

The little boy's head was hatless,
And tears in each little eye.

"Why don't you go home to your momma?" I said
And this was the children's reply

"Our momma's in jail, they locked her up.
"Left Jim and I alone,
"So we've come here to sleep in the tents tonight,
"For we have no mother, no home.

"Our papa got hurt in the shooting Friday night,
"We waited all night for him,
"For he was a union guard you know,
"But he never came home any more."

THE BIG FAT BOSS AND THE WORKERS—*Ella May (1929)*

The song is played to the tune of "Polly Wolly Doodle" and appears in *Let's Stand Together*.

The boss man wants our labor, and money to packaway,
The workers wants a union and the eight-hour day.

The boss man hates the workers, the workers hates the boss,
The boss man rides in a big fine car, and the workers has to walk.

The boss man sleeps in a big fine bed, and dreams of his silver and gold,
The workers sleeps in an old straw bed and shivers from the cold.

Fred Beal he is in prison, a-sleeping on the floor,
But he will soon be free again, and speak to us some more.

The union is a-growing, the ILD is strong,
We're going to show the bosses that we have starved too long.

MILL MOTHER'S LAMENT—*Ella May (1929)*

Music is from the ballad "Little Mary Phagan." The song is preserved in Margaret Larkin's *Ella May Wiggins and Songs of the Gastonia Textile Strike* and was originally published in the International Labor Defense's October 1929 *Labor Defender*.

We leave our homes in the morning,
We kiss our children good bye,
While we slave for the bosses,
Our children scream and cry.

And when we draw our money,
Our grocery bills to pay,
Not a cent to spend for clothing,
Not a cent to lay away.

And on that very evening,
Our little son will say,
"I need some shoes, Mother,
"And so does sister May."

How it grieves the heart of a mother,
You everyone must know,
But we can't buy for our children,
Our wages are too low.

It is for our little children,
That seems to us so dear,
But for us nor them, dear workers,
The bosses do not care.

But understand, all workers,
Our union they do fear,
Let's stand together, workers,
And have a union here.

UP IN OLD LORAY—*Odell Corley (1929)*

The verses are sung to the popular song "On Top of Old Smokey."
It is found in *American Folksongs of Protest*. It was originally published
in the August 1929 issue of *Labor Defender*.

Up in old Loray,
Six stories high,
That's where they found us,
Ready to die.

CHORUS: Go pull off your aprons,
                Come join our strike.
                Say "Good-bye old bosses,
                "We're going on strike."

The bosses will starve you,
They'll tell you more lies,
Than there's crossties on the railroads,
Or stars in the skies.

The bosses will rob you,
They will take half you make,

And claim that you took it up,
In coupon books.

Up in old Loray,
All covered with lint,
That's where our shoulders
Was crippled and bent.

Up in old Loray,
All covered with cotton,
It will carry you to your grave,
And you soon will be rotten.

LET ME SLEEP IN YOUR TENT TONIGHT, BEAL— Odell Corley (1929)

This is a parody of "Let Me Sleep in Your Barn Tonight, Mister." It is found in *American Folksongs of Protest*.

Let me sleep in your tent tonight, Beal,
For it's cold lying out on the ground,
And the cold wind is whistling around us,
And we have no place to lie down.

Manville-Jenckes has done us dirty,
And has set us out on the ground.
We are sorry we did not join you,
When the rest walked out and joined.

Oh, Beal, please forgive us,
And take us into your tent,
We will always stick to the union,
And not scab on you no more.

You have tore up our hall and you wrecked it,
And you've went and threw out our grub,
Only God in his heaven,
Knows what you scabs done to us.

COME ON YOU SCABS IF YOU WANT TO HEAR— Odell Corley (1929)

This ballad is set to the tune of "Casey Jones." It is found in both Margaret Larkin's *Ella May Wiggins and Songs of the Gastonia Textile Strike* and *North Carolina Folklore*, Volume II.

Come on you scabs if you want to hear.
The story of a cruel millionaire,
Manville Jenckens was the millionaire's name,
He bought the law with his money and frame, (frame up)
But he can't buy the union with his money and frame.

Told Violet Jones if she'd go back to work,
He'd buy her a new Ford and pay her well for her work,
They throwed rotten eggs at Vera and Beal on the stand,
They caught the man with the pistol in his hand,
Trying to shoot Beal on the speaking stand.

They took Beal to the Monroe jail,
They put him in a dirty cell,
But Beal and the strikers put up a darn good fight,
We'll make the bosses howl and hear old Manville say,
"It ain't no use fighting the union this way."

I BOUGHT A SCAB FOR FIFTY CENTS—*Odell Corley* (1929)

This ballad was sung to the tune of "Mademoiselle from Armentieres." This song is found in *Strike Songs of the Depression*.

> I bought a scab for fifty cents, parlie voo,
> I bought a scab for fifty cents, parlie voo,
> I bought a scab for fifty cents,
> The son of a gun jumped over the fence.
> Hinky Dinky, parlie voo.

THE SPEAKERS DIDN'T MIND—*Daisy McDonald* (1929)

This is set to the music of "The Wreck of the Old '97." This ballad is printed in *American Folksongs of Protest* and *Strike Songs of the Depression*.

> On a summer night as the speaking went on,
> All the strikers were satisfied,
> The thugs threw rotten eggs at the speakers on the stand,
> It caused such a terrible fright.

> The speakers didn't mind that and spoke right on,
> As speakers want to do,

It wasn't long till the police came,
To shoot them through and through.

On that very same night the mob came down,
To the union ground you know,
Searching high and low for the boys and men,
Saying "Damn you, come on, let's go.

"We'll take you to jail and lock you up,
"If you're guilty or not we don't care,
"Come git out of these tents, you low down dogs,
"Or we'll kill you all right here."

They arrested the men, left the women alone,
To do the best they can,
They tore down their tents, run them out in the woods,
"If you starve we don't give a damn."

Our poor little children, they had no homes,
They were left in the streets to roam,
But the W.I.R. put up more tents and said,
"Children, come on back home."

Some of our leaders are already free,
Hoping all the rest will be soon,
And if they do we'll yell with glee,
For the South will be on a boom.

Fread Beal and Sophie and all the rest,
Are our best friends, we know,
For they come to the South to organize,
When no one else would go.

They've been our friends and let's be theirs,
And help them organize,
We'll have more money and better homes,
And live much better lives.

On A Summer Eve—Daisy McDonald (1929)

Daisy again used "The Wreck of the Old '97" for the tune. It is found in *American Folksongs of Protest* and *Strike Songs of the Depression*. It was originally published in the November 1929 edition of *Labor Defender*.

On a summer eve as the sun was setting,
And the wind blew soft and dry,

They locked up all our union leaders,
While tears stood in our eyes.

Fred Beal's in jail with many others,
Facing the electric chair,
But we are working with the ILD,
To set our leaders clear.

Come on fellow workers and join the union,
Also the ILD,
Come help us fight this great battle,
And set our leaders free.

Come listen, fellow workers, about poor Ella May,
She lost her life on the state highway,
She'd been to a meeting as you all can see,
Doing her bit to get our leaders free.

She left five children in this world to roam,
But the ILD gave them a brand new home,
So workers come listen and you will see,
It pays all workers to join the ILD.

If we love our brothers as we all should do,
We'll join this union help fight it through,
We all know the boss don't care if we live or die,
He'd rather see us hang on the gallows high.

Our leaders in prison are our greatest friends,
But the ILD will fight to the end,
Come on fellow workers, join the ILD,
And do your part to set our leaders free.

We need them back on the firing line,
To carry on the work that they left behind,
When they were put in the dirty cell,
In the Gastonia jail we all know well.

Barbara Allen

This popular ballad is over four centuries old and multiple versions of it are found throughout the United States of America and the world. This rendition from Lila Ripley Barnwell of Henderson, North Carolina, was published in Charlotte's *Charlotte Observer* in 1913. This version is found in *North Carolina Folklore*, Volume II. © 1952, Duke University Press. All rights reserved. Republished by permission of the copyright holder. www.dukeupress.edu.

In Scarlet Town where I was born
There was a fair maid dwelling
Made every youth cry "Well-away"
Her name was Barbara Allen.

All in the merry month of May
When green buds they were swelling,
Young Jimmy Grove on his death bed lay
For the love of Barbara Allen.

He sent his man unto her then,
To the town where she was dwellin';
"You must come to my master dear," he said,
"If your name is Barbara Allen."

So slowly, slowly she came up
And slowly she came nigh him.
And all she said as there she stood:
"Young man, I think you're dying."

He turned his face unto the wall,
And death was with him dealing.
"Adieu, adieu, my dear friends all,
Adieu to Barbara Allen."

As she was walkin o'er the fields
She spied the corpse a-coming.
"Lay down. Lay down the corpse," she said,
"That I may look upond him."

With scornful eye she looked down,
Her cheeks with laughter swellin'.
And all her friends cried out amain,
"Oh, shameful Barbara Allen!"

When he was dead and laid in grave
Her heart was struck with sorrow.
"Oh, mother, mother, make my bed,
"For I shall die tomorrow.

"Farewell," she said, "ye virgins all,
"And shun the fault I fell in;
"Henceforth take warning by the fall
"Of cruel Barbara Allen."

RED RIVER VALLEY

This version of this classic love/loss ballad is from Minnie Church in Heaton, North Carolina, from 1930. It is published in *North Carolina*

> From this valley they tell me you are going.
> How I'll miss your blue eyes and bright smiles!
> For you carry with you all the sunshine
> That has brightened my path for a while.
>
> CHORUS: Let's consider a while ere you leave me
> Do not hasten to bid me adieu,
> But remember the bright Red River Valley
> And the girl who has loved you so true.
>
> I have waited a long time, my darling,
> For the word you never would say,
> But alas, my poor heart it is breaking,
> For they tell me you are going away.
>
> Do you think of the move you are leaving,
> How lonely and dreary it will be:
> Do you think of the fond heart you are breaking
> And the girl who has loved you so true?

BALLAD OF ELLA MAE—*Malvina Reynolds (1955)*

The music in this ballad is original. It was printed in Margaret Larkin's *Ella May Wiggins and Songs of the Gastonia Textile Strike*. It was released in 1955 and 1983 and copyright is held by the Schroder Music Company.

> I never could do for my children,
> Not even keep them alive, it seems,
> Though they came of this small body,
> And they built me, all of my dreams.
>
> CHORUS: And if you want to know why I'm for the union,
> And walking this picket line,
> It's the one best thing that I ever could do,
> For those babies of mine.
>
> She never had much schooling,
> Because she went to the mills so young,

But the songs came like cool spring water,
To Ella Mae's mountain tongue.

Four babies were sick with fever,
They needed their mama so bad,
But Ella Mae stayed on at the mill,
To earn them doctor and bread.

Oh please let me work the day shift,
Oh change my run if you might,
My little ones sick of the fever,
They cry for me in the night.

But supers are stupid and cruel,
The Company's answer is No,
And Ella came home with the last paycheck.
And four of her babes had to go.

The Company killed Ella's children,
And they feared the power of her song,
So they shot Ella Mae on Gastonia's streets,
And they never have paid for their wrong.

Gastonia's unions are growing,
The workers are stronger each day,
And the voice that sounds clearest among them all,
Is the singing of young Ella Mae.

THE MURDER OF ELLA MAY WIGGINS SONG— *Gary Green (1976)*

Gary Green released the song on his Folkways recording, *These Six Strings Neutralize the Tools of Oppression*. It was rereleased in 2012 on Smithsonian Folkways Recordings.

© 1976 and 2012 Gary Green as recorded on Smithsonian Folkways Records

Come with me to Loray Mill in Gastonia, North Carolina
To a woman and her story that history forgot to tell.
Ella May Wiggins lived alone with her nine little children
Each one she loved oh so well.
She had to work graveyard shift to tend them in the day;
Hard work and little pay.
CHORUS: A hobo must ramble; a cowboy must ride
　　　　　Every train has tracks upon which it must glide

> While some will choose the mountain crest,
> Some will choose the shore;
> But some will jump the track and say, "I'll run for you no more."

Then four of her babies come down with whooping cough
And they needed their mother in the nights.
She asked the supervisor to let her work the day.
He said "NO," so she left her job at Loray.
Then her four babies died… Though Ella May had tried;
What's her power against the super at Loray?

Now Ella May told and sang her sad story.
She said, "We'll always be slaves unless we organize."
Gaston County needs a union of the workers…
Then, our voices they would recognize.

Soon all Gaston workers were talkin' union…
And old Loray had a strike upon its hands.
Ella May led every union meetin' singin' her songs to clappin' hands.
So the newspaper and the big mill bosses said, "This is getting out of hand."
And on the fourteenth day of September, they put a bullet through Ella May's brain.

Chapter Notes

Introduction

1. Vera Buch Weisbord, "GASTONIA, 1929 Strike at the Loray Mill," 190; Jacquelyn Dowd Hall, "Disorderly Women: Gender and Labor Militancy in the Appalachian South," 355; Alan Lomax, Woody Guthrie, and Pete Seeger, *Hard Hitting Songs for Hard-Hit People*, 180; *Gastonia Gazette*, September 19, 1929.

2. *Evening Times*, May 1985.

3. Rita Ann Jackson, interview by author, January 30, 2004, and February 26, 2004; *Gaston Gazette*, November 6, 2011.

4. Darlene Horton, interview by author, December 20, 2000; *Star-Gazette*, December 12, 1977.

5. Horton, interview by author, December 20, 2000.

6. Jo Lynn Haessly, "Mill Mother's Lament: Ella May, Working Women's Militancy, and the 1929 Gaston County Strikes" (master's thesis, University of North Carolina, 1987), 28, 29; *Charlotte Observer*, September 1979.

7. Weisbord, "GASTONIA, 1929 Strike at the Loray Mill," 203.

8. John A. Salmond, *Gastonia 1929: The Story of the Loray Mill Strike* (Chapel Hill: University of North Carolina Press, 1995), 127.

9. Robert L. Williams and Elizabeth W. Williams, *The Thirteenth Juror* (Kings Mountain, NC: Herald House, 1983), 152, 153; Robert L. Williams, *People Worth Meeting and Stories Worth Repeating* (Dallas, NC: Southeastern, 2000), 49; *Star-Gazette*, December 12, 1977; Salmond, *Gastonia 1929*, 166.

10. Williams, *The Thirteenth Juror*, 152, 153.

11. Cook, "Spinning Through Time."

Prologue

1. Lomax, Guthrie, Seeger, *Hard Hitting Songs for Hard-Hit People*, 180; Weisbord, "Gastonia," 190; *Gastonia Gazette*, September 19, 1929; Hall, "Disorderly Women," 355.

2. Draper, "Gastonia Revisited", 16–17; Larkin, "Tragedy in North Carolina", 686; Williams, *The Thirteenth Juror*, xiii.

Chapter 1

1. Bobby Bolch, interview by author, March 7, 2004; Haessly, "Mill Mother's Lament," 24; *Gastonia Gazette,* September 16, 1929; Barium Springs Home for the Children Papers, September 17, 1929–December 27, 1934.

2. Mellinger Edward Henry, *Folk-Songs from the Southern Highlands* (New York: J.J. Augustin, 1938), 1.

3. Margaret Morley, *The Carolina Mountains* (Boston: Houghton Mifflin, 1913), 85, 86; Millie Wandell, interview by Donald Fredrick, November 24, 1984.

4. Henry, *Folk Songs from the Southern Highlands*, 10, 11.

5. Henry, *Folk Songs from the Southern Highlands*, 11.

6. Morley, *The Carolina Mountains*, 110, 112, 113, 125.

7. Morley, *The Carolina Mountains*, 86, 128.

8. Morley, *The Carolina Mountains*, 113.

9. Bob Pickelsimer, Genealogy research on May family, received February 2004; Haessly, "Mill Mother's Lament," 1; Patrick Huber, "Mill Mother's Lament: Ella May Wiggins and the Gastonia Textile Strike of 1929," *Southern Cultures Journal* 15, no. 3 (Fall 2009): 83; Certificate of Death: Ella May Wiggins, North Carolina Vital Records, Gaston County, September 1929.

10. Pickelsimer, Genealogy; Certificate of Death: Lewis Jackson May, North Carolina Vital Records, Cherokee County, July 1914.

11. Pickelsimer, Genealogy; Certificate of Death: Lewis Jackson May; Barium Springs Home for the Children Papers, September 17, 1929–December 27, 1934; Huber, "Mill Mother's Lament," 83.

12. Haessly, "Mill Mother's Lament," 1, 2; Florence Cope Bush, *Dorie: Woman of the Mountains* (Knoxville: University of Tennessee Press, 1992), 77.

13. Bush, *Dorie*, 83, 96; Haessly, "Mill Mother's Lament," 2.

14. Haessly, "Mill Mother's Lament," 2.

15. Bush, *Dorie*, 83, 96; Huber, "Mill Mother's Lament," 83.

16. Bush, *Dorie*, 92; Haessly, "Mill Mother's Lament," 3.

17. Bush, *Dorie*, 115, 116.

18. Haessly, "Mill Mother's Lament," 3; John Greenway, *American Folksongs of Protest* (Philadelphia: University of Pennsylvania Press, 1953), 245.

19. Haessly, "Mill Mother's Lament," 3, 4; Department of Commerce, *Fourteenth Census of the United States 1920-Population*.

20. Haessly, "Mill Mother's Lament," 3; Pickelsimer, Genealogy; Huber, "Mill Mother's Lament," 83.

21. Haessly, "Mill Mother's Lament," 5; Herron, interview by author, March 7, 2004.

22. Haessly, "Mill Mother's Lament," 2, 3, 4; Certificate of Birth: Clyde Francis Wiggins, North Carolina Birth Index and Vital Statistics, Cherokee, NC.

23. Haessly, "Mill Mother's Lament," 6.

24. Haessly, "Mill Mother's Lament," 4.

25. Haessly, "Mill Mother's Lament," 5, 6; Greenway, *American Folksongs of Protest*, 245.

26. Haessly, "Mill Mother's Lament," 7.

27. Haessly, "Mill Mother's Lament," 7; Certificate of Birth: Millie Wiggins.

Chapter 2

1. Haessly, "Mill Mother's Lament," 7; Certificate of Death: Guy Wesley Wiggins. North Carolina Vital Records, Gaston County, June 1926.

2. Jacquelyn Dowd Hall, Robert Korstad, and James Leloudis, "Cotton Mill People: Work, Community, and Protest in the Textile South, 1880–1940," *American History Review* 91, no. 2 (April 1986): 262.

3. Salmond, *Gastonia 1929*, 11; Ron Cook, *Spinning Through Time: Gaston County and the Textile Industry* (Charlotte, NC: WTVI, 1996).

4. Salmond, *Gastonia 1929*, 10; Cook, *Spinning Through Time*.

5. Salmond, *Gastonia 1929*, 10; Robin Hood, "The Loray Mill Strike" (master's thesis, University of North Carolina, 1932), 9, 12; Cook, *Spinning Through Time*.

6. Cook, *Spinning Through Time*.

7. Haessly, "Mill Mother's Lament," 7; Hood, "The Loray Mill Strike," 12; Jacquelyn Dowd Hall, James Leloudis, Robert Korstad, Mary Murphy, Lu Ann Jones, and Christopher B. Daly, *Like a Family: The Making of a Southern Cotton Mill World* (Chapel Hill: University of North Carolina Press, 1987), 214.

8. Lois MacDonald, *Southern Mill Hills: A Study of Social and Economic Forces in Certain Textile Mill Villages* (New York: Alex L. Hillman, 1928), 10, 11; Cook, *Spinning Through Time*.

9. MacDonald, *Southern Mill Hills*, 10, 11; Tom Hanchett and Ryan Summer, *Images of America: Charlotte and the Carolina Piedmont* (Mt. Pleasant, SC: Levine Museum of the New South, Arcadia, 2003), 25; Cook, *Spinning Through Time*.

10. MacDonald, *Southern Mill Hills*, 6, 7; Cook, *Spinning Through Time*.

11. *Gastonia Gazette*, April 5, 1929.

12. Salmond, *Gastonia 1929*, 2; Hall, Korstad and Leloudis, "Cotton Mill People," 247.

13. Rosalyn Baxandall, Linda Gordan, and Susan Reverby, *America's Working Women: A Documentary History 1600 to the Present* (New York: Vintage Books, Random House, 1976), 262; Vera Buch Weisbord, *A Radical Life* (Bloomington: Indiana University Press, 1977), 170; Salmond, *Gastonia 1929*, 2.

14. Cook, *Spinning Through Time*; Hall, Korstad and Leloudis, "Cotton Mill People," 265.

15. Cook, *Spinning Through Time*; Margaret Larkin, "Tragedy in North Carolina," *North American Review* (December 1929): 687; Hall, Korstad and Leloudis, "Cotton Mill People," 265.

16. Fred Beal, *Word from Nowhere: The Story of a Fugitive from Two Worlds* (Great Britain: Billing and Sons, Ltd., Guildford and Esher, 1938), 131; Weisbord, *A Radical Life*, 184.

17. Haessly, "Mill Mother's Lament," 11–13.

18. Ibid., 14.

19. Ibid., 15; Hall, Korstad and Leloudis, "Cotton Mill People," 265.

20. Baxandall, Gordon and Reverby, *America's Working Women*, 262; Weisbord, *A Radical Life*, 170.

21. Baxandall, Gordon and Reverby, *America's Working Women*, 262; Weisbord, *A Radical Life*, 170.

22. Haessly, "Mill Mother's Lament," 17; Dr. Jennings J. Rhyne, *Some Southern Cotton Mill Workers and Their Villages* (Chapel Hill: University of North Carolina Press, 1930), 2.

23. Charlotte Crawford, Hilda Gunst, Enid Kiefer, Elizabeth Thornburg and others, *Bessemer City Centennial 1893—1993* (Charlotte, NC: Walsworth, 1993), 49.

24. Glenda Elizabeth Gilmore, *Defying Dixie: The Radical Roots of Civil Rights 1919—1950* (New York: W.W. Norton, 2008), 87; Weisbord, "Gastonia, 1929 Strike at the Loray Mill," 186.

25. Cook, *Spinning Through Time*;

Theodore Draper, "Gastonia Revisited," *Social Research* 38, no. 1 (Spring 1971): 4; Hanchett and Sumner, *Images of America*, 54.

26. Gilmore, *Defying Dixie*, 75; Salmond, *Gastonia 1929*, 51.

27. Haessly, "Mill Mother's Lament," 28.

28. Gilmore, *Defying Dixie*, 75, 76; Haessly, "Mill Mother's Lament," 25.

29. Colleen Hall, interview by author, December 20, 2000; Horton, interview by author, December 20, 2000; Haessly, "Mill Mother's Lament," 25.

30. Hall, interview by author, December 20, 2000; Wandell, interview by Fredrick, November 24, 1984.

31. Hall, interview by author, December 20, 2000; Wandell, interview by Fredrick, November 24, 1984; Haessly "Mill Mother's Lament," 18, 19; Certificate of Death: Guy Wiggins.

32. Haessly, "Mill Mother's Lament," 22–24.

33. Haessly, "Mill Mother's Lament," 23; Weisbord, "GASTONIA, 1929 Strike at the Loray Mill," 190.

34. Haessly, "Mill Mother's Lament," 23.

35. Haessly, "Mill Mother's Lament," 23.

36. Haessly, "Mill Mother's Lament," 18–19; Horton, interview by author, December 20, 2000; Wandell, Fredrick interview, November 24, 1984.

37. Weisbord, *A Radical Life*, 185.

38. Bush, *Dorie*, 218; Haessly, "Mill Mother's Lament," 18; Barium Springs Home for the Children Papers, September 17, 1929–December 27, 1934.

39. Horton, interview by author, December 20, 2000; Larkin, "Tragedy in North Carolina," 686; Haessly, "Mill Mother's Lament," 29.

40. Horton, interview by author, December 20, 2000; Wandell, interview with author, December 26, 2000; Haessly, "Mill Mother's Lament," 29.

41. Horton, interview by author, December 20, 2000; Haessly, "Mill Mother's Lament," 28–30.

42. *New York Times*, April 11, 1929.

43. Mary Frederickson and Joyce Kornbluh, *Sisterhood and Solidarity: Workers'*

Education for Women, 1914–1984 (Philadelphia: Temple University Press, 1984), 24; J.G. Van Osdell, Jr., "The Textile Strike of 1929 and Its Repercussions" (master's thesis, Tulane University, 1962), 7.

44. Haessly, "Mill Mother's Lament," 26–28.

45. Robert Allison Ragan, "LORAY MILLS Gastonia NC 1900," in *The Pioneer Cotton Mills of Gaston County (NC) "The First Thirty" (1848–1904) and Gaston Textile Pioneers* (Charlotte, NC: Ragan and Company, 1973), 1–3.

46. Ragan, "LORAY MILLS Gastonia NC 1900," 1–2.

47. Broadus Mitchell, *The Rise of Cotton Mills in the South*, 2nd ed. (Columbia: University of South Carolina, 2001), 109; Ellen Grigsby, "The Politics of Protest: Theoretical, Historical and Literary Perspectives on Labor Conflict in Gaston County, North Carolina" (Ph.D. thesis, University of North Carolina, 1986), 103.

48. Cook, *Spinning Through Time*; Draper, "Gastonia Revisited," 4–5.

49. Van Osdell Jr., "The Textile Strike of 1929 and its Repercussions," 5; Cook, *Spinning Through Time*.

Chapter 3

1. Hall et al., *Like a Family*, 213; Hall, "Disorderly Women," 354–355.

2. Larkin, "Tragedy in North Carolina," 687; Weisbord, "GASTONIA, 1929 Strike at the Loray Mill," 186, 188.

3. Draper, "Gastonia Revisited," 6.

4. K.O. Byers, "My Life Story," *Labor Defender* (August 1929): 155.

5. Draper, "Gastonia Revisited," 5.

6. Weisbord, "GASTONIA, 1929 Strike at the Loray Mill," 193; Salmond, *Gastonia 1929*, 15.

7. Baxandall, Gordon and Reverby, *America's Working Women*, 262.

8. Baxandall, Gordon and Reverby, *America's Working Women*, 262; Williams and Williams, *The Thirteenth Juror*, 29; Larkin, "Tragedy in North Carolina," 686.

9. Mary Jo Buhle, Paul Buhle, and Dan Georgakas, *Encyclopedia of the American Left* (New York: Oxford University Press, 1998), 254; Gilmore, *Defying Dixie*, 69.

10. Buhle, Buhle, and Georgakas, *Encyclopedia of the American Left*, 254; Gilmore, *Defying Dixie*, 69.

11. Williams and Williams, *The Thirteenth Juror*, 25, 26, 28.

12. *Gastonia Gazette*, April 3, 1929; Weisbord, "GASTONIA, 1929 Strike at the Loray Mill," 188; Weisbord, *A Radical Life*, 104; Draper, "Gastonia Revisited," 9, 10.

13. Fred Beal, *The Red Fraud: An Exposure of Stalinism* (New York: Tempo, 1949), 9; Salmond, *Gastonia 1929*, 18; William Dunne, *Gastonia: Citadel of the Class Struggle in the New South* (New York: Workers Library, 1929), 19; Gilmore, *Defying Dixie*, 75; *TIME*, May 31, 1948.

14. Williams and Williams, *The Thirteenth Juror*, 25; Salmond, *Gastonia 1929*, 18; Draper, "Gastonia Revisited," 10.

15. Williams and Williams, *The Thirteenth Juror*, 30.

16. Beal, *Word from Nowhere*, 92, 93, 96, 97.

17. Ibid., 100.

18. Ibid., 92, 93; Hall, Korstad, and Leloudis, "Cotton Mill People," 275.

19. Beal, *Word from Nowhere*, 100; Salmond, *Gastonia 1929*, 19; Thomas Parramore, "The Hundred Days of Ella May," in *Carolina Quest* (Englewood Cliffs, NJ: Prentice-Hall, 1978), 369.

20. Beal, *Word from Nowhere*, 100; Weisbord, "GASTONIA, 1929 Strike at the Loray Mill," 194.

21. Beal, *Word from Nowhere*, 100; Weisbord, "GASTONIA, 1929 Strike at the Loray Mill," 194.

22. Weisbord, *A Radical Life*, 179.

23. Williams and Williams, *The Thirteenth Juror*, 51.

24. Beal, *Word from Nowhere*, 102, 106; Weisbord, *A Radical Life*, 178.

25. Beal, *Word from Nowhere*, 102.

26. Ibid., 103, 104; Weisbord, *A Radical Life*, 178.

27. Salmond, *Gastonia 1929*, 20; Beal, *Word from Nowhere*, 104, 105.

28. Salmond, *Gastonia 1929*, 20, 21; Weisbord, *A Radical Life*, 178, 179; Beal, *Word from Nowhere*, 104, 105.

29. Beal, *Word from Nowhere*, 109.

30. Ibid., 109, 110; Salmond, *Gastonia 1929*, 23, 24; *Gastonia Gazette*, April 2.

31. Beal, *Word from Nowhere*, 110, 111; Weisbord, *A Radical Life*, 178; *Gastonia Gazette*, April 3, 1929.

32. *Gastonia Gazette*, April 3, 1929.

33. Ibid.

Chapter 4

1. *Gastonia Gazette*, April 2, 3, 1929; Beal, *Word from Nowhere*, 112.

2. *Gastonia Gazette*, April 3, 1929.

3. Ibid., April 2, 3; Salmond, *Gastonia 1929*, 24; Beal, *Word from Nowhere*, 112.

4. Gilmore, *Defying Dixie*, 77; Beal, *Word from Nowhere*, 112; Salmond, *Gastonia 1929*, 28.

5. *Gastonia Gazette*, April 4, 1929; Salmond, *Gastonia 1929*, 28; Beal, *Word from Nowhere*, 112, 113; Draper, "Gastonia Revisited," 11.

6. Beal, *Word from Nowhere*, 112, 113.

7. Beal, *The Red Fraud*, 13.

8. Draper, "Gastonia Revisited," 15, 16; Buhle, Buhle and Georgakas, *Encyclopedia of the American Left*, 255.

9. Beal, *The Red Fraud*, 13.

10. Beal, *Word from Nowhere*, 112, 116.

11. Weisbord, "GASTONIA, 1929 Strike at the Loray Mill," 188, 191; Dee Garrison, *Mary Heaton Vorse: The Life of an American Insurgent* (Philadelphia: Temple University, 1989), 218.

12. Buhle, Buhle and Georgakas, *Encyclopedia of the American Left*, 255; Beal, *Word from Nowhere*, 112.

13. Beal, *Word from Nowhere*, 118, 119.

14. Ibid., 118; Weisbord, *A Radical Life*, 176.

15. *Gastonia Gazette*, April 5, 1929.

16. Ibid.

17. Weisbord, *A Radical Life*, 182.

18. Hood, "The Loray Mill Strike," 37.

19. *Gastonia Gazette*, April 3, 1929.

20. Ibid., April 4, 1929.

21. Ibid., April 4, 1929.

22. Ibid., April 5, 1929.

23. Ibid.

24. Beal, *Word from Nowhere*, 122, 123.

25. Frederickson and Kornbluh, *Sisterhood and Solidarity*, 175.

26. Salmond, *Gastonia 1929*, 30.

27. Salmond, *Gastonia 1929*, 30.

28. Salmond, *Gastonia 1929*, 30.

29. Wandell, Fredrick interview, November 24, 2004.

30. *Gastonia Gazette*, April 3, 1929; *New York Times*, April 4, 1929.

31. *Gastonia Gazette*, April 3, 4, 1929; *New York Times*, April 4, 1929.

32. *Gastonia Gazette*, April 3, 4, 1929; *New York Times*, April 4, 1929.

33. Weisbord, *A Radical Life*, 177.

34. *New York Times*, April 4.

35. *Gastonia Gazette*, April 4; Weisbord, *A Radical Life*, 176.

36. Margaret Larkin, "Ella May's Songs," *Nation* 129, no. 3353 (October 9, 1929): 382, 384; Weisbord, "GASTONIA, 1929 Strike at the Loray Mill," 190; Weisbord, *A Radical Life*, 185.

37. Eugene Feldman, "Looking Back Ella May Wiggins: The Songs of Struggle," *KEEP STRONG* (August 1979): 36, 37.

38. Weisbord, "GASTONIA, 1929 Strike at the Loray Mill," 191.

39. Carl Reeve, "The Great Gastonia Textile Strike," *Political Affairs* 63 (March 1984): 37–39.

40. Ibid.

41. Ibid., 39, 40.

42. Williams, *The Thirteenth Juror*, 52.

43. Buch, *A Radical Life*, 194; Beal, *Word from Nowhere*, 124, 125.

44. Beal, *Word from Nowhere*, 125.

45. Karl Reeve, "Gastonia Sees and Learns," *Labor Defender* (June 1929): 117.

46. Rev. James Myers, *Field Notes on Textile Strikers in the South*, Ministers Meeting in Gastonia, NC, 1929.

47. *Gastonia Gazette*, September 16, 1929; Salmond, *Gastonia 1929*, 43, 44.

48. Salmond, *Gastonia 1929*, 37, 38, 43–45.

49. *Gastonia Gazette*, April 16, 17, 1929; Gilmore, *Defying Dixie*, 87.

50. *Gastonia Gazette*, April 17, 1929.

51. *Gastonia Gazette*, April 17, 24, 1929.

52. *Gastonia Gazette*, April 19, 1929; *New York Times*, April 19, 1929.

53. *Gastonia Gazette*, April 19, 1929; *New York Times*, April 19, 1929.

54. *Gastonia Gazette*, April 20, 1929; Salmond, *Gastonia 1929*, 38, 45; *New York Times*, April 23, 1929.

55. *Gastonia Gazette*, April 20, 1929; Atkins Papers, Stephen Dolley Letter.

56. Carl Reeve, "Gastonia: The Strike, The Frameup, the Heritage," *Political Affairs* 63 (April 1984), 23.

57. *New York Times*, April 23, 1929; *Gastonia Gazette*, April 23, 1929.

58. *Gastonia Gazette*, April 24, 25, 26, 1929.

59. Ibid., April 23, 1929.

60. Ibid., April 15, 19, 24, 1929; Reeve, "Gastonia: The Strike, The Frameup, the Heritage," 26.

61. Weisbord, *A Radical Life*, 205; Gilmore, *Defying Dixie*, 87.

62. Weisbord, *A Radical Life*, 205; *Gastonia Gazette*, April 29, 1929.

63. *Let's Stand Together: The Story of Ella Mae Wiggins* (Charlotte, NC: National Organization for Women Metrolina Chapter, September 14, 1979), 8; *Star-Gazette*, December 12, 1977.

64. Beal, *Word from Nowhere*, 123; Weisbord, "GASTONIA, 1929 Strike at the Loray Mill," 194; Margaret Supplee Smith, and Emily Herring Wilson, *North Carolina Women Making History* (Chapel Hill and London: University of North Carolina Press, 1999), 264; Wandell, interview by Fredrick, November 24, 1984.

65. *Let's Stand Together*, 9; Weisbord, "GASTONIA, 1929 Strike at the Loray Mill," 194.

66. Wandell, interview by Fredrick, November 24, 1984.

67. Gilmore, *Defying Dixie*, 29; Reeve, "Gastonia: The Strike, The Frameup, the Heritage," 28.

68. Edward L. Ayers, *The Promise of the New South: Life After Reconstruction* (New York: Oxford University Press, 1992), vii, viii, ix, x, 3.

69. Ibid., 3, 151, 153.

70. Smith and Wilson, *North Carolina Women Making History*, 265; Weisbord, "GASTONIA, 1929 Strike at the Loray Mill," 188, 194; Reeve, "Gastonia: The Strike, The Frameup, the Heritage," 28.

71. *Gastonia Gazette*, May 11, 1929.

72. Weisbord, "GASTONIA, 1929 Strike at the Loray Mill," 194.

73. Ibid.

74. Haessly, "Mill Mother's Lament," 28.

75. Weisbord, "GASTONIA, 1929 Strike at the Loray Mill," 194, 195.

76. Grigsby, "The Politics of Protest," 204.

77. Dunne, *Gastonia*, 27.

78. Ibid., 29; Salmond, *Gastonia 1929*, 66.

79. Gilmore, *Defying Dixie*, 67, 70, 79.

80. Ibid., 90; Otto Hall, "Gastonia and the Negro," *Labor Defender* (August 1929): 153; Salmond, *Gastonia 1929*, 66, 67; Garrison, *Mary Heaton Vorse*, 220.

81. Salmond, *Gastonia 1929*, 58; Hood, "The Loray Mill Strike," 98; *Gastonia Gazette*, May 3, 10, 1929; Reeve, "Gastonia: The Strike, the Frameup, and the Heritage," 26, 27.

82. Salmond, *Gastonia 1929*, 58.

83. Reeve, "Gastonia: The Strike, the Frameup, and the Heritage," 26; Salmond, *Gastonia 1929*, 58; Weisbord, *A Radical Life*, 185; *Gastonia Gazette*, May 11, 1929; Hood, "The Loray Mill Strike," 99.

84. Haessly, "Mill Mother's Lament," 99, 100; *Gastonia Gazette*, May 10, 1929.

85. *Gastonia Gazette*, May 3, 8, 10, 1929; Haessly, "Mill Mother's Lament," 100.

86. Reeve, "Gastonia: The Strike, the Frameup, and the Heritage," 27.

87. Salmond, *Gastonia 1929*, 58; *Gastonia Gazette*, May 13, 1929.

88. Salmond, *Gastonia 1929*, 58, 59; Reeve, "Gastonia: The Strike, the Frameup, and the Heritage," 27; *Gastonia Gazette*, May 3, 1929.

89. Reeve, "Gastonia: The Strike, the Frameup, and the Heritage," 28; Salmond, *Gastonia 1929*, 59.

90. Reeve, "Gastonia: The Strike, the Frameup, and the Heritage," 28.

91. Weisbord, "GASTONIA, 1929 Strike at the Loray Mill," 189.

92. Baxandall, Gordan, Reverby, *America's Working Women*, xiv.

93. Ibid.; Salmond, *Gastonia 1929*, 31; *Gastonia Gazette*, April 4, 1929.

94. Salmond, *Gastonia 1929*, 60; Larkin, "Tragedy in North Carolina," 690.

95. Salmond, *Gastonia 1929*, 56, 60; "Pioneers Return from Soviet Union," *Labor Defender* (November 1929): 221.

96. "Pioneers Return from Soviet Union," 221.

97. "Who Are the Gastonia Prisoners?" *Labor Defender* (September 1929): 171–172; Weisbord, *A Radical Life*, 212.

98. Weisbord, *A Radical Life*, 212, 216, 217; Salmond, *Gastonia 1929*, 60.

99. "Who Are the Gastonia Prisoners?" 171–172; Weisbord, "GASTONIA, 1929 Strike at the Loray Mill," 200.

100. Salmond, *Gastonia 1929*, 20, 45; Beal, *Word from Nowhere*, 104, 105.

101. Weisbord, "GASTONIA, 1929 Strike at the Loray Mill," 187.

102. Ibid.; "Who Are the Gastonia Prisoners?" 171–172; Beal, *Word from Nowhere*, 115.

103. *Gastonia Gazette*, May 6, 8, 1929; Salmond, *Gastonia 1929*, 56.

104. Beal, *Word from Nowhere*, 126–128.

105. Horton, interview by author, December 20, 2000; Haessly, "Mill Mother's Lament," 108.

106. Jackson, interview by author, February 26, 2004; Haessly, "Mill Mother's Lament," 108.

107. Beal, *Word from Nowhere*, 134, 135; Jackson, interview by author, February 26, 2004.

108. Beal, *The Red Fraud*, 15.

109. *Let's Stand Together*, 11, 12.

110. Salmond, *Gastonia 1929*, 70; *Gastonia Gazette*, August 14, 1929.

111. Salmond, *Gastonia 1929*, 71.

112. *Gastonia Gazette*, June 8, 1929; Weisbord, "GASTONIA, 1929 Strike at the Loray Mill," 197; Beal, *The Red Fraud*, 17.

113. *Gastonia Gazette*, June 8, 10, 1929.

114. Haessly, "Mill Mother's Lament," 109.

115. Mildred Gwen Andrews, *The Men and the Mills: A History of the Southern Textile Industry* (Macon, GA: Mercer University Press, 1987), 89; American Civil Liberties Union, "Justice—North Carolina Style" (New York: ACLU, May 1930), 5; *Gastonia Gazette*, June 8, 10, 1929; Weisbord, "GASTONIA, 1929 Strike at the Loray Mill," 197, 198.

116. Beal, *Word from Nowhere*, 143, 144.

117. *Gastonia Gazette*, June 8, 10, 1929.

118. *Gastonia Gazette*, June 10, 1929.

119. *Gastonia Gazette*, June 15, 1929; Haessly, "Mill Mother's Lament," 112.

120. Reeve, "The Great Gastonia Textile Strike," 37; *Gastonia Gazette*, August 28, 1929; Beal, *Word from Nowhere*, 146.

121. Haessly, "Mill Mother's Lament," 112; *Gastonia Gazette*, June 19, 1929.

122. *Gastonia Gazette*, June 20, 1929.

123. *Gastonia Gazette*, June 24, 1929; Haessly, "Mill Mother's Lament," 111, 112.

124. Greenway, *American Folksongs of Protest*, 134.

125. Ibid., 134, 135.

126. Williams and Williams, *The Thirteenth Juror*, Preface.

127. Ibid., 26; Salmond, *Gastonia 1929*, 84, 87, 93, 142.

128. *Gastonia Gazette*, August 29, 1929; *New York Times*, June 24, 1929; Reeve, "The Great Gastonia Textile Strike," 40; Salmond, *Gastonia 1929*, 84.

129. Salmond, *Gastonia 1929*, 114; Beal, *Word from Nowhere*, 153; *Gastonia Gazette*, August 28, 1929.

130. Williams and Williams, *The Thirteenth Juror*, 129, 132.

131. *Gastonia Gazette*, August 29, 1929; Salmond, *Gastonia 1929*, 113, 115–117.

132. *Gastonia Gazette*, August 29, 1929; Salmond, *Gastonia 1929*, 56, 116, 117.

133. Salmond, *Gastonia 1929*, 86, 87; Dunne, *Gastonia*, 8.

134. Dunne, *Gastonia*, 46.

135. Salmond, *Gastonia 1929*, 117; Williams, *The Thirteenth Juror*, 129, 133.

136. Salmond, *Gastonia 1929*, 118, 119.

137. Williams, *The Thirteenth Juror*, 129; ACLU, "Justice—North Carolina Style," 6.

138. Nell Battle Lewis, "Anarchy vs.

Communism in Gastonia," *Nation* (September 25, 1929): 321, 322.

139. Ibid., 321; ACLU, "Justice—North Carolina Style," 6, 7.

140. Ella May, "What I Believe," *Labor Defender* (November 1929): 227.

141. *Gastonia Gazette*, September 11, 1929.

142. Liston Pope, *Millhands and Preachers: A Study of Gastonia* (New Haven, CT: Yale University Press, 1942), 292.

143. Ibid.

Chapter 5

1. Haessly, "Mill Mother's Lament," 3.

2. Newman White, *North Carolina Folklore*, volume II (The Frank C. Brown Collection, Durham, NC: Duke University Press, 1952), 598–600; Haessly, "Mill Mother's Lament," 3.

3. *Star-Gazette*, December 12, 1977.

4. Timothy P. Lynch, *Strike Songs of the Depression* (Jackson: University Press of Mississippi, 2001), 25.

5. Lynch, *Strike Songs of the Depression*, 25; Weisbord, "GASTONIA, 1929 Strike at the Loray Mill," 192.

6. Weisbord, "GASTONIA, 1929 Strike at the Loray Mill," 192.

7. Archie Green, *Wobblies, Pile Butts, and Other Heroes: Laborlore Explorations* (Urbana: University of Illinois Press, 1993), 277–278.

8. Green, *Wobblies*, 278–279.

9. Green, *Wobblies*, 278; Lynch, *Strike Songs of the Depression*, 22–23.

10. Green, *Wobblies*, 278; Lynch, *Strike Songs of the Depression*, 22–23.

11. Lynch, *Strike Songs of the Depression*, 27–29.

12. Ibid., 29–30.

13. Larkin, "Ella May's Songs," 384.

14. *Let's Stand Together*, 8.

15. Weisbord, *A Radical Life*, 260; *Evening Times*, May 1985.

16. White, *North Carolina Folklore*, volume II, 498, 499.

17. Gilmore, *Defying Dixie*, 93; Larkin "Ella May's Songs," 382.

18. Atkins Papers.

19. Greenway, *American Folksongs of Protest*, 249, 250; *Let's Stand Together*, 20.

20. Weisbord, "GASTONIA, 1929 Strike at the Loray Mill," 199, 200; Lomax, Guthrie, Seeger, *Hard Hitting Songs for Hard-Hit People*, 185.

21. *Let's Stand Together*, 22.

22. Ibid.

23. *Let's Stand Together*, 18; Larkin, "Ella May Wiggins and Songs of the Gastonia Textile Strike," 9, 10.

24. Greenway, *American Folksongs of Protest*, 135, 136.

25. Greenway, *American Folksongs of Protest*, 137, 138.

26. Larkin, "Ella May Wiggins and Songs of the Gastonia Textile Strike," 8; White, *North Carolina Folklore*, volume II, 510, 511; Larkin, "Ella May's Songs," 385; Lynch, *Strike Songs of the Depression*, 33.

27. Ibid., 41–43.

28. Larkin, "Tragedy in North Carolina," 690.

29. Greenway, *American Folksongs of Protest*, 138, 139; Lynch, *Strike Songs of the Depression*, 43, 44.

30. Weisbord, "GASTONIA, 1929 Strike at the Loray Mill," 199.

31. Larkin, "Ella May Wiggins and Songs of the Gastonia Textile Strike," 34, 35.

32. Green, correspondence with author, December 11, 2012; Lomax, Guthrie, and Seeger, *Hard Hitting Songs for Hard-Hit People*, 182.

33. Green, correspondence with author, December 11, 2012.

34. Gary Green, "The Murder of Ella May Wiggins," in *These Six Strings Neutralize The Tools of Oppression* (New York: Folkways Records, 1977), 3.

35. Beal, *Word from Nowhere*, 131; Weisbord, *A Radical Life*, 184.

Chapter 6

1. *Gastonia Gazette*, September 16, 1929; *New York Times*, September 15, 1929.

2. *Gastonia Gazette*, September 14, 1929.

3. *Gastonia Gazette*, September 10, 11, 14, 1929.

4. Salmond, *Gastonia 1929*, 127.

5. *New York Times*, September 15, 1929.

6. *Gastonia Gazette*, September 16, 1929; Beal, *Word from Nowhere*, 160; *Spartanburg Herald*, September 16, 1929.

7. *Gastonia Gazette*, September 16, 1929, March 7, 1930.

8. *Gastonia Gazette*, September 16, 1929, March 3, 1930; Certificate of Death: Ella May Wiggins, September 1929.

9. *Let's Stand Together*, 13.

10. *Gastonia Gazette*, February 26, 1930.

11. *Gastonia Gazette*, September 16, 1929, February 26, 1930; *Spartanburg Herald*, September 16, 1929.

12. *Star-Gazette*, December 12, 1977; Wandell, Fredrick interview, November 24, 1984.

13. *Star-Gazette*, December 12, 1977; Wandell, Fredrick interview, November 24, 1984; Jackson, interview by author, February, 26, 2004; *New York Times*, September 16, 1929.

14. *New York Times*, September 15, 1929.

15. *Gastonia Gazette*, September 19, 1929.

16. Ibid.

17. Michael W. Frye, "The Loray Strike Viewed by Six North Carolina Newspapers" (master's thesis, University of North Carolina, 1972), 44, 46, 47.

18. *Let's Stand Together*, 14.

Chapter 7

1. Margaret Larkin, "The Story of Ella May," *New Masses* 5, no. 6 (November 1929): 4.

2. Pope, *Millhands and Preachers*, 294, 295; Haessly, "Mill Mother's Lament," 116; *Gastonia Gazette*, September 16, 1929, November 6, 2011.

3. Larkin, "Ella May's Songs," 385.

4. Ibid.; Salmond, *Gastonia 1929*, 133.

5. *Gastonia Gazette*, September 17, 1929; Haessly, "Mill Mother's Lament," 116.

6. Haessly, "Mill Mother's Lament," 116; *Gastonia Gazette*, September 17, 23, 1929.

7. *Gaston Gazette*, November 6, 2011; Barium Springs Home for the Children Papers, September 17, 1929–December 27, 1934; *Charlotte Observer*, September 1979.

8. Ibid.

9. Jackson, interview by author, February 26, 2004; Bolch, interview by author, March 7, 2004; Wandell, interview by author, December 26, 2000; *Charlotte Observer*, September 1979; Payne, interview by Fredrick November 24, 1984; Wandell, interview by Fredrick, November 24, 1984.

10. *Gastonia Gazette*, September 17, 18, 1929; Draper, "Gastonia Revisited," 19.

11. *Gastonia Gazette*, September 19, 1929; Pope, *Millhands and Preachers*, 293; Van Osdell, Jr., "The Textile Strike of 1929 and its Repercussions," 21; ACLU, "Justice—North Carolina Style," 7.

12. Ibid.

13. *Gastonia Gazette*, May 3, 1930; Draper, "Gastonia Revisited," 27.

14. Joseph Morrison, *Governor O. Max Gardner: A Power in North Carolina and New Deal Washington* (Chapel Hill: University of North Carolina Press, 1971), 61, 62.

15. ACLU, "Justice—North Carolina Style," 3.

16. *Gastonia Gazette*, April 19, 1929; Lynch, *Strike Songs of the Depression*, 32, 33; Jones, e-mail correspondence to author, March 21, 2004; *New York Times*, September 16, 1929.

17. Salmond, *Gastonia 1929*, 128, 129, 155; *New York Times*, September 16, 1929, *Gastonia Gazette*, April 19, October 22, 1929; Wandell, interview by Fredrick, November 24, 1984.

18. *Gastonia Gazette*, October 23, 1929.

19. *Gastonia Gazette*, October 23, 1929, June 13, 2004; *Raleigh News and Observer*, October 25, 1929; Salmond, *Gastonia 1929*, 155.

20. Salmond, *Gastonia 1929*, 155; Tom Tippett, *When Southern Labor Stirs* (Binghamton, NY: Vail-Ballou Press, 1931), 107; *Gastonia Gazette*, October 23, 1929.

21. Beal, *The Red Fraud*, 36; Beal, *Word from Nowhere*, 174; Williams and Williams, *The Thirteenth Juror*, 43; Lynch, *Strike Songs of the Depression*, 40.

22. Draper, "Gastonia Revisited," 27.

23. *TIME*, May 31, 1948; *Gaston Gazette*, November 6, 2011.

24. Salmond, *Gastonia 1929*, 177, 178; Weisbord "Gastonia, 1929 Strike at the Loray Mill," 203; Weisbord, *A Radical Life*, 261; Wandell, interview by Fredrick, November 24, 1984.

25. *Gastonia Gazette*, October 26, 1929; ACLU, "Justice—North Carolina Style," 7.

26. *Gastonia Gazette*, February 24, March 6, 1930.

27. *Gastonia Gazette*, February 24, March 7, 1930.

28. Ibid.

29. *Gastonia Gazette*, February 28, March 1, 1930.

30. *Gastonia Gazette*, March 4–6, 1930.

31. Ibid.

32. *Gastonia Gazette*, March 5, 6, 1930.

33. *Gastonia Gazette*, February 25, March 3, 1930; Salmond, *Gastonia 1929*, 158, 159, 163.

34. *Gastonia Gazette*, February 26, 1930; Salmond, *Gastonia 1929*, 164.

35. *Gastonia Gazette*, March 7, 1930.

36. Ibid.

37. ACLU, "Justice—North Carolina Style," 9, 10; Haessly, "Mill Mother's Lament," 116, 117.

38. Hall, Korstad, and Leloudis, "Cotton Mill People," 278, 280; Timothy Ellis, *A Centennial History of Loray Baptist Church: Gastonia, North Carolina 1905–2005, Lighting the Way for 100 Years* (Spartanburg, South Carolina: Trenton Creative Enterprises, 2005), 97.

39. Tippett, *When Southern Labor Stirs*, 108.

40. Haessly, "Mill Mother's Lament," 21.

41. *Gastonia Gazette*, April 11, 1930.

42. Crawford, Gunst, Kiefer, Thornburg, *Bessemer City Centennial 1893–1993*, 87; Draper, "Gastonia Revisited," 25, 26; Robert Allison Ragan, *The Textile Heritage of Gaston County, North Carolina, 1848–2000: One Hundred Mills and the Men Who Built Them* (Charlotte, NC: Ragan and Company, 2001), 116.

43. Gilmore, *Defying Dixie*, 96, 98; *Gastonia Gazette*, March 7, 1930.

44. Gilmore, *Defying Dixie*, 106–108, 112.

45. Salmond, *Gastonia 1929*, 175; *Gastonia Gazette*, March 8, 1930.

46. Gilmore, *Defying Dixie*, 119–122

47. Hall, Korstad and Leloudis, "Cotton Mill People," 276, 277; Hanchett and Sumner, *Images of America*, 91; Salmond, *Gastonia 1929*, 181.

48. Haessly, "Mill Mother's Lament," 138; Draper, "Gastonia Revisited," 26; Hall, Korstad and Leloudis, "Cotton Mill People," 280, 281, 283; Hanchett and Sumner, *Images of America*, 53.

49. Beal, *The Red Fraud*, 11; Larkin, "Tragedy in North Carolina," 689.

50. Haessly, "Mill Mother's Lament," 136, 137.

51. Baxandall, Gordan, and Reverby, *America's Working Women*, 263; Hall, Korstad, and Leloudis, "Cotton Mill People," 284, 285.

52. Cook, *Spinning Through Time*; *Gaston Gazette*, November 6, 2011.

53. Cook, *Spinning Through Time*.

54. *Star-Gazette*, December 12, 1977.

55. Ibid.

56. Ibid.

57. Horton, interview by author, December 20, 2000; *Charlotte Observer*, September 1979.

58. *Gaston Gazette*, "Ella Mae Wiggins Honored," September 1979.

59. Wandell, interview by Fredrick, November 24, 1984.

60. Ibid.; *Gaston Gazette*, "Ella Mae Wiggins Honored," September 1979; *Let's Stand Together*, 3; *Charlotte Observer*, "50 Bitter Years," September 1979.

61. "History Happened Here," June 12, 2004; *Gaston Gazette*, June 13, 2004.

62. *Gaston Gazette*, June 13, 2004; Huber, "Mill Mother's Lament," 106.

63. Crawford, Gunst, Kiefer, Thornburg, *Bessemer City Centennial 1893–1993*, 49. Ragan, *The Textile Heritage of Gaston County*, 116.

64. *Gaston Gazette*, 1986.

65. Huber, "Mill Mother's Lament," 106.

66. Ibid., 104; *Evening Times*, May 85.

67. Buhle, Buhle, and Georgakas, *Encyclopedia of the American Left*, 255, 625; Haessly, "Mill Mother's Lament," intro.

Bibliography

American Civil Liberties Union. "Justice—North Carolina Style." New York: ACLU, May 1930.

Andrews, Mildred Gwen. *The Men and the Mills: A History of the Southern Textile Industry.* Macon, GA: Mercer University Press, 1987.

Atkins Papers. Bessemer City Worker Letter. Records of J.W. Atkins, Managing Editor of *Gastonia Gazette* in 1929. Possession of Claire Atkins Pittman.

Atkins Papers. Stephen Dolley Letter. Records of J.W. Atkins, Managing Editor of *Gastonia Gazette* in 1929. Possession of Claire Atkins Pittman.

Ayers, Edward L. *The Promise of the New South: Life After Reconstruction.* New York: Oxford University Press, 1992.

"Babies in the Mill," "Textile Voices." *Sing Out!* 29, no. 3, Carry It On.

Barium Springs Home for the Children Papers. Papers include report card, facts for identification, and medical records. September 17, 1929–December 27, 1934.

Baxandall, Rosalyn, Linda Gordan, and Susan Reverby. *America's Working Women: A Documentary History 1600 to the Present.* New York: Vintage Books, Random House, 1976.

Beal, Fred. *The Red Fraud: An Exposure of Stalinism.* New York: Tempo, 1949.

_____. *Word from Nowhere: The Story of a Fugitive from Two Worlds.* Great Britain: Billing and Sons, Ltd., Guildford and Esher, 1938.

Bolch, Bobby. Interview by author. Gastonia, NC. March 7, 2004.

Buhle, Mary Jo, Paul Buhle, and Dan Georgakas. *Encyclopedia of the American Left.* New York: Oxford University Press, 1998.

Bush, Florence Cope. *Dorie: Woman of the Mountains.* Knoxville: University of Tennessee Press, 1992.

Byers, K.O. "My Life Story." *Labor Defender* (August 1929): 155.

Certificate of Birth: Clyde Francis Wiggins. North Carolina Birth Index and Vital Statistics. Cherokee, NC.

Certificate of Birth: Millie Wiggins. South Carolina Department of Health and Environmental Control. Division of Vital Records. Columbia, SC.

Certificate of Death: Ella May Wiggins. North Carolina Vital Records, Gaston County. September 1929.

Certificate of Death: Guy Wesley Wiggins. North Carolina Vital Records, Gaston County. June 1926.

Certificate of Death: Lewis Jackson May. North Carolina Vital Records. Cherokee County. July 1914.

Charlotte Observer (Charlotte, North Carolina).

Cook, Ron. *Spinning Through Time: Gaston County and the Textile Industry.* Charlotte, NC: WTVI, 1996.

Crawford, Charlotte, Hilda Gunst, Enid Kiefer, Elizabeth Thornburg and others. *Bessemer City Centennial 1893—1993.* Charlotte, NC: Walsworth, 1993.

Department of Commerce, Bureau of the Census. *Fourteenth Census of the United States 1920-Population.* Swain County, North Carolina. January 26,

1920. North Carolina Room. Charlotte/Mecklenburg County Library, North Carolina.

Draper, Theodore. "Gastonia Revisited." *Social Research* 38, no. 1 (Spring 1971): 3–29.

Dunne, William. *Gastonia: Citadel of the Class Struggle in the New South.* New York: Workers Library, 1929.

Ellis, Timothy. *A Centennial History of Loray Baptist Church: Gastonia, North Carolina 1905–2005, Lighting the Way for 100 Years.* Spartanburg, South Carolina: Trenton Creative Enterprises, 2005.

Evening Times (Sayre, Pennsylvania).

Feldman, Eugene. "Looking Back—Ella May Wiggins: The Songs of Struggle." *KEEP STRONG* (August 1979): 36–37.

"50,000 New Members in the ILD." *Labor Defender* (November 1929): 226.

Fowke, Edith, and Joe Glazer. *Songs of Work and Protest.* New York: Dover, 1973.

Frederickson, Mary, and Joyce Kornbluh. *Sisterhood and Solidarity: Workers' Education for Women, 1914–1984.* Philadelphia: Temple University Press, 1984.

Frye, Michael W. "The Loray Strike Viewed by Six North Carolina Newspapers." Master's thesis, University of North Carolina, 1972.

Garrison, Dee. *Mary Heaton Vorse: The Life of an American Insurgent.* Philadelphia: Temple University, 1989.

Gaston Gazette (Gastonia, North Carolina).

Gastonia Gazette. (Gastonia, North Carolina).

Gilmore, Glenda Elizabeth. *Defying Dixie: The Radical Roots of Civil Rights 1919–1950.* New York: W.W. Norton, 2008.

Green, Archie. *Wobblies, Pile Butts, and Other Heroes: Laborlore Explorations.* Urbana: University of Illinois Press, 1993.

Green, Gary. "The Murder of Ella May Wiggins." In *These Six Strings Neutralize The Tools of Oppression.* New York: Folkways Records, 1977.

Greenway, John. *American Folksongs of Protest.* Philadelphia: University of Pennsylvania Press, 1953.

Grigsby, Ellen. "The Politics of Protest: Theoretical, Historical and Literary Perspectives on Labor Conflict in Gaston County, North Carolina." Ph.D. thesis, University of North Carolina, 1986.

Haessly, Jo Lynn. "Mill Mother's Lament: Ella May, Working Women's Militancy, and the 1929 Gaston County Strikes." Master's thesis, University of North Carolina, 1987.

Hall, Colleen. Interview by author. December 20, 2000.

Hall, Jacquelyn Dowd. "Disorderly Women: Gender and Labor Militancy in the Appalachian South." *Journal of American History* 73 (September 1986): 354–382.

Hall, Jacquelyn Dowd, Robert Korstad, and James Leloudis. "Cotton Mill People: Work, Community, and Protest in the Textile South, 1880–1940." *American History Review* 91, no. 2 (April 1986): 245–286.

Hall, Jacquelyn Dowd, James Leloudis, Robert Korstad, Mary Murphy, Lu Ann Jones, and Christopher B. Daly. *Like a Family: The Making of a Southern Cotton Mill World.* Chapel Hill: University of North Carolina Press, 1987.

Hall, Otto. "Gastonia and the Negro." *Labor Defender* (August 1929): 153.

Hanchett, Tom, and Ryan Sumner. *Images of America: Charlotte and the Carolina Piedmont.* Mt. Pleasant, SC: Levine Museum of the New South, Arcadia, 2003.

Henry, Mellinger Edward. *Folk-Songs from the Southern Highlands.* New York: J.J. Augustin, 1938.

Herron, Lorrie Absher. Interview by author. March 7, 2004.

"History Happened Here: Many Stories

of Loray." Gastonia: North Carolina Humanities Council, Weaving Cultures and Communities, June 12, 2004.

Hood, Robin. "The Loray Mill Strike." Master's thesis, University of North Carolina, 1932.

Horton, Darlene. Interview by author. December 20, 2000.

Huber, Patrick. *Linthead Stomp*. Chapel Hill: University of North Carolina Press, 2008.

_____. "Mill Mother's Lament: Ella May Wiggins and the Gastonia Textile Strike of 1929." *Southern Cultures Journal* 15, no. 3 (Fall 2009): 81–110.

Jackson, Mark Allan. *Prophet Singer*. Jackson: University Press of Mississippi, 2007.

Jackson, Rita Ann. Phone interview by author. January 30, 2004, and February 26, 2004.

Larkin, Margaret. "Ella May's Songs." *Nation* 129, no. 3353 (October 9, 1929): 382–85.

_____. "Ella May Wiggins and Songs of the Gastonia Textile Strike." Charlotte, NC: 1929.

_____. "The Story of Ella May." *New Masses* 5, no. 6 (November 1929): 3–5.

_____. "Tragedy in North Carolina." *North American Review* (December 1929): 686–90.

Let's Stand Together: The Story of Ella Mae Wiggins. Charlotte, NC: National Organization for Women Metrolina Chapter, September 14, 1979.

Lewis, Nell Battle. "Anarchy vs. Communism in Gastonia." *Nation* (September 25, 1929): 321–22.

Lomax, Alan, Woody Guthrie, and Pete Seeger. *Hard Hitting Songs for Hard-Hit People*. New York: Oak, 1967.

Lynch, Timothy P. *Strike Songs of the Depression*. Jackson: University Press of Mississippi, 2001.

MacDonald, Lois. *Southern Mill Hills: A Study of Social and Economic Forces in Certain Textile Mill Villages*. New York: Alex L. Hillman , 1928.

May, Ella. "What I Believe." *Labor Defender* (November 1929): 227.

Mitchell, Broadus. *The Rise of Cotton Mills in the South*. 2nd ed. Columbia: University of South Carolina, 2001.

Morley, Margaret. *The Carolina Mountains*. Boston: Houghton Mifflin, 1913.

Morrison, Joseph. *Governor O. Max Gardner: A Power in North Carolina and New Deal Washington*. Chapel Hill: University of North Carolina Press, 1971.

Myers, the Rev. James. *Field Notes on Textile Strikes in the South*. Ministers Meeting in Gastonia, NC, 1929.

New York Times (New York, New York).

News and Observer (Raleigh, North Carolina).

Parramore, Thomas. "The Hundred Days of Ella May." In *Carolina Quest*. Englewood Cliffs, NJ: Prentice-Hall, 1978.

Payne, Charlotte. Interview by Donald Fredrick. November 24, 1984.

Pickelsimer, Bob. Genealogy research on May family. Received February 2004.

"Pioneers Return from Soviet Union." *Labor Defender* (November 1929): 221.

Pope, Liston. *Millhands and Preachers: A Study of Gastonia*. New Haven, CT: Yale University Press, 1942.

Ragan, Robert Allison. "LORAY MILLS Gastonia NC 1900." In *The Pioneer Cotton Mills of Gaston County (NC) "The First Thirty" (1848–1904) and Gaston Textile Pioneers*. Charlotte, NC: Ragan and Company, 1973.

_____. *The Textile Heritage of Gaston County, North Carolina, 1848–2000: One Hundred Mills and the Men Who Built Them*. Charlotte, NC: Ragan and Company, 2001.

Reeve, Carl. "Gastonia: The Strike, the Frameup, and the Heritage." *Political Affairs* 63 (April 1984).

_____. "The Great Gastonia Textile Strike." *Political Affairs* 63 (March 1984).

Reeve, Karl. "Gastonia Sees and Learns." *Labor Defender* (June 1929): 117.

Bibliography

Rhyne, Dr. Jennings J. *Some Southern Cotton Mill Workers and Their Villages.* Chapel Hill: University of North Carolina Press, 1930.

Seeger, Pete. *American Industrial Ballads.* Smithsonian/Folkways Recordings CD. 1992. Originally issued in 1957.

Salmond, John A. *Gastonia 1929: The Story of the Loray Mill Strike.* Chapel Hill: University of North Carolina Press, 1995.

Smith, Margaret Supplee, and Emily Herring Wilson. *North Carolina Women Making History.* Chapel Hill and London: University of North Carolina Press, 1999.

"Solidarity Forever." *New York Teacher* 42, no. 6 (November 22, 2000).

Star-Gazette (Elmira, New York).

TIME. May 31, 1948.

Tippett, Tom. *When Southern Labor Stirs.* Binghamton, NY: Vail-Ballou Press, 1931.

Van Osdell Jr., J.G. "The Textile Strike of 1929 and its Repercussions." Master's thesis, Tulane University, 1962.

Wandell, Candy. Interview by author. December 26, 2000.

Wandell, Millie. Interview by Donald Fredrick. November 24, 1984.

Weisbord, Vera Buch. "GASTONIA, 1929 Strike at the Loray Mill." *Southern Exposure* 1, nos. 3 & 4 (Winter 1974): pp. 185–203.

_____. *A Radical Life.* Bloomington: Indiana University Press, 1977.

White, Newman. *North Carolina Folklore,* Volume II. The Frank C. Brown Collection. Durham, NC: Duke University Press, 1952.

_____. *North Carolina Folklore,* Volume III. The Frank C. Brown Collection. Durham, NC: Duke University Press, 1952.

"Who Are the Gastonia Prisoners?" *Labor Defender* (September 1929): 171–72.

Williams, Robert L. *People Worth Meeting and Stories Worth Repeating.* Dallas, NC: Southeastern, 2000.

Williams, Robert L., and Elizabeth W. Williams. *The Thirteenth Juror.* Kings Mountain, NC: Herald House, 1983.

Index